The ESSENCE of GASTRONOMY

Understanding the Flavor of Foods and Beverages

The ESSENCE
of GASTRONOMY

Understanding the Flavor of Foods and Beverages

Peter Klosse

CRC Press
Taylor & Francis Group
Boca Raton London New York

CRC Press is an imprint of the
Taylor & Francis Group, an **informa** business

CRC Press
Taylor & Francis Group
6000 Broken Sound Parkway NW, Suite 300
Boca Raton, FL 33487-2742

Version Date: 20131031

International Standard Book Number-13: 978-1-4822-1676-9 (Hardback)

Visit the Taylor & Francis Web site at
http://www.taylorandfrancis.com

and the CRC Press Web site at
http://www.crcpress.com

CONTENTS

Preface   xiii
Acknowledgments   xvii
Prologue   xix

1 Introduction   1

There Is No Reason to Argue about Taste   1
Struggling with Words   2
Source of Pleasure   4
Times Have Changed   6
Setting the Stage   8
Gastronomy, the Science of Flavor and Tasting   10
Chemical Senses and Classification   12
Universal Flavor Factors   14
    Mouthfeel   15
    Flavor Richness   15
Gastronomy and Hotel Schools   16
Structure of This Book   17
Summary   18
Notes   19

2 Making Sense of Taste   21

Tasting: An Intricate Process   22
Tasting: What Happens in the Mouth?   23
The Gustatory System   25
    Papillae, Taste Buds, and Taste Cells   25

Personal Differences   27
Basic Tastes: Problems and Misconceptions   30
The Olfactory System   33
    The Olfactory Process   33
    Odor Composition and Preferences   35
    Olfaction, Age, and Health   36
The Trigeminal System   37
    No Flavor without Some Kind of Mouthfeel   37
    The Receptors of the Trigeminal System   38
    The Versatility of Nociceptors   40
    Astringency and Trigeminal Sensitivity   41
There Is More than Meets the Mouth   42
    Sensory Thresholds and Time Intensity   42
    Adaptation and (De)Sensitization   43
    Flavor Enhancement or Suppression   44
    Carry-Over Effects   46
Intrinsic and Extrinsic Components of Flavor   48
    Synesthesia and Other Forms of Sensory Interaction   50
Are Professionals Able to Taste Better than Amateurs?   51
Summary   53
Notes   54

3   Understanding Flavor   61

Finding Words: Comparing Apples to Oranges   61
Words Do Not Come Easily   63
Universal Flavor Factors   64
    Mouthfeel: Types   64
    Flavor Intensity   68
    The Flavor Type (Fresh or Ripe Flavor Tones)   69
    Flavor Richness and Complexity   70
Building the Theoretical Model   72
Description of the Flavor Styles   76
    Flavor Styles 1 and 5, Neutral and Robust   76
    Flavor Styles 2 and 6, Round and Full   78
    Flavor Styles 3 and 7, Fresh and Pungent   78
    Flavor Styles 4 and 8, Balanced, from Low to High   78
How to Generate a Flavor Profile   79

Flavor Profiles Are Dynamic   82
The Role of the Basic Flavors   84
  Sweet   84
  Salt   86
  Sour   87
  Bitter   88
  Umami   90
  The Basic Flavors from a Different Perspective   93
The Role of the Big Four Basic Food Molecules   93
  Water/Moisture   93
  Lipids: Fats and Oils   95
  Carbohydrates   98
  Proteins   100
Summary   104
Appendix: How Low-Fat Foods Get Their Texture   106
Notes   107

4 Flavor in the Kitchen   **111**

No Beast Is a Cook   111
Practice or Purpose?   112
Ingredients: Basic Qualities   115
  The Influence of Climate and Location on Flavor   118
  The Influence of Cultivars, Varietals, and Breeds on Flavor   119
  The Influence of Production Methods on Flavor   121
  The Influence of Ripeness on Flavor   123
The Essence of Cooking   125
Characteristics of the Principal Techniques   128
  Cooking and Blanching   128
  Poaching   129
  Steaming   130
  Frying/Pan Frying/Stir Frying (Bao Technique)   131
  Sautéing/Stir Frying (Chao Technique)   132
  Roasting/Baking   133
  Grilling/Barbecuing   134
  Simmering/Braising/Stewing   135
  Deep-Frying/Shallow Frying   136
Other Techniques to Get or to Keep Flavor   137

Searing, Papillotte, Clay Pot, *Croûte, Sous-Vide*   137
Cold Cooking: Brining, Curing, Smoking   138
Drying, Pickling, Marinating   140
Very Cold Cooking: Liquid Nitrogen and Other Molecular
Novelties   141
Cooking Times and the Influence of Altitude and Pressure   143
A Microwave Is Not an Oven   144
Making the Dish Complete: Sauces, Herbs and Spices,
Vegetables, Fruits, and Mushrooms   146
The Role and Types of Sauces   147
The Role of Herbs and Spices   149
Vegetables   150
Fruits   153
Mushrooms   155
Seasonality of Vegetables and Fruits   156
Flavor Composition   159
Technique Follows Function   159
The Arithmetic of Flavor and the Art of Omission   160
Details Matter   161
Presentation and Foodpairing   162
Culinary Success Factors   163
Palatability and Liking   163
Factor 1: Name and Presentation Fit the Expectation   165
Factor 2: Appetizing Smell That Fits the Food   165
Factor 3: Good Balance in Flavor Components in Relation
to the Food   166
Factor 4: Presence of Umami   166
Factor 5: Combination of Hard and Soft Textures   167
Factor 6: High Flavor Richness   168
Palatability Objectified   169
Summary   170
Appendix: Influence of Preparation   172
Notes   174

5   **The Flavor of Beverages**   **177**

Aqueous Solutions   177
Mineral Water or Spring Water   178

Mineral Water and the Interaction with Wine and Food   180
Tea   183
Flavor Styles of Tea   184
Other Uses of Tea   187
Coffee   188
From Plant to Bean   188
From Bean to Cup   190
Beer   192
The Process of Brewing   193
Beer Types and Flavor Profiles   196
Wine   198
Demystification Required   199
Focus on Flavor   201
The Effects of Grapes and Grape Growing on the
Flavor of Wine   202
Terroir, Climate, and Grape Variety   202
Agricultural Choices and Practices   204
Harvest   207
The Effects of Vinification on the Flavor of Wine   210
The Role of the Skins   210
The Role of Yeast   211
The Role of Temperature   213
The Role of the Lees   214
The Role of Malolactic Fermentation   216
The Role of Wood   217
After Vinification   220
Blending   220
Further Aging   222
Marketing (Bottle, Label, Price, Distribution)   223
Serving   225
Beer and Wine Comparison   226
Summary   228
Notes   229

6   Matching Foods and Beverages: The Fundamentals   231

Taste!   231
Old Guidelines   233

New Guidelines   234
    The Dominant Factor   235
Gustatory Effects in Matching Foods and Beverages   237
    The Rice Test   237
    Acid Observations   239
    Sweet Connections   242
    Flavor Styles and Descriptions   243
Mouthfeel Effects in Matching Foods and Beverages   245
    Types of Coating   245
    Fizzy Secrets   246
    The Dry Run   247
    Spicy Problems   249
    Say Cheese   250
Gastronomic Reflections   254
    Harmony and Contrast   255
    Learning to Like   256
    Cultural Context   257
    Cooking toward Wine and Precision   258
    Expensive Wines   260
    Marriages Made in Heaven   261
    Regional Wines with Regional Foods and Other Customs   262
    Robust and Elegant   263
    Beer and Food   265
Summary   267
Notes   268

7   **Food Appreciation and Liking**   **271**

Introduction   271
Tasting and the Brain   272
    From Receptor to the Brain   272
    Synthesizing Neural Information   276
    Cross-Modal Interaction   278
Food Liking and Palatability   280
    We Like What We Need   280
    We Like What We Do Not Need   282
    Liking and the Gut   284
    Development of Preferences   285

Expectancy and Liking 287
External Influences on Liking 287
Palatability and the Aesthetics of Gastronomy 289
Flavor Styles Revisited 291
The Theorem of Flavor Styles 293
The New Paradigm 294
Is Food the Stimulus or the Response? 296
Where It All Comes Together: The Behavioral Model
of Food Choice 298
Introducing: Guestronomy 300
Summary 302
Notes 303

Epilogue 307

Index 311

PREFACE

MY ADVENTUROUS AND REWARDING
JOURNEY IN GASTRONOMY

I cannot guarantee that this book will blow your mind—whoever could you give such a guarantee? However, I can assure you that this book will give you a view on gastronomy that is likely to be new. It may even astound you. This view and the subsequent theory have slowly developed and matured for almost 25 years. It started with wine research in 1988 and evolved into a successful school, training sommeliers, producing books, a PhD, and ultimately a chair in gastronomy both at Stenden University in Leeuwarden and at the Hotel Management School of Maastricht. This all happened in the Netherlands.

It is all in the genes—apparently. In 1955, my parents opened a restaurant at a tourist attraction (De Echoput), near Apeldoorn in the Netherlands. It is beautifully situated amidst the Crown forests, and the restaurant's history is a tale in itself. It grew from a simple, family type restaurant into one of the cornerstones of Dutch gastronomy. The restaurant is now housed in a five-star hotel that has been developed at the same location and is still operated by the family. The kitchen is renowned for the preparation of products from the forest: game, berries, and mushrooms. Such a local perspective is popular today but was nothing new to us. We have not done anything else since the early 1960s.

Wine has also been a focus of interest since the early days. It has become a family tradition to handle our own imports. It gave me the ultimate chance to be in contact with wine makers. I wanted to know the person behind the wine; what motivates him, what his goals are, and what he needs to do to accomplish those goals. Good friendships were the result, which helped me to better understand the world of wine and what it takes to make something special.

In a good restaurant, sensible advice on the combination of wines and dishes is expected. As a gastronome by birth, I am aware of the importance of this advice. Wines and dishes can strongly influence each other, and not all matches are successful. I wanted to know why, to learn about the fundamentals of gastronomy, and to be able to predict if a certain combination could be successful.

Therefore, starting in 1988, I enthusiastically hosted this fundamental research focused on the flavor of wine. During an extended period of time, a group led by scientific researchers met on a regular basis for tastings to describe the differences between wines. Gradually, it yielded all kinds of useful, practical information, not specifically about the flavor of wine but on flavor in general.

It marked the beginning of an interesting period in which many different things started to become clear—a period of searching for fundamental answers, of testing an evolving theory and trying to validate it. This process had all the pitfalls and deceptions you might expect. We experienced how a beautiful, ripe, and voluptuous Corton Charlemagne of a good producer turned into a horrific acidic beverage when served with a spicy risotto and prawns. And also how another white Burgundy, too old and forgotten and by far not as exclusive as the Corton Charlemagne, magically turned into something great with a roast leg of lamb—completely unexpectedly. These and many more experiences welded together this new theory on flavor. A theory that we were not looking for but that came to us.

The theory on gastronomy and the restaurant progressed simultaneously and organically, not purposely. One step naturally led to the next one. Nevertheless, it would not have taken place without another family trait: dare to be—and to think—independently and to always ask why. My father dared to change the traditional preparation of game. He abolished marinades, heavy sauces, and garnishes. He treated game as a meat from nature and prepared it and served it as naturally as possible. It took quite some convincing, but he succeeded. The world of gastronomy is also full of traditional theories and opinions that are hardly ever seriously questioned. I hope to convince you of the benefits of this other way of looking at gastronomy. I have succeeded in doing this in my own country, and it is time to take the next step—to show the world that the new flavor theory is powerful in addressing all kinds of gastronomic issues. It provides answers to and explanations for questions that have long seemed too difficult to even consider.

There is one thing, though, that this book asks from you. Open-mindedness proved to be indispensable in building this theory. So it will help when you, the reader, bring an open mind to considering this new approach. Relevant words need to be redefined, concepts reconsidered, and strong beliefs reassessed. Your willingness to read this book with a free spirit will be rewarded. The rest of your life will benefit from understanding gastronomy and all related concepts better. You will receive more enjoyment from pairing wines and foods. You will find new combinations and will be better at making those combinations.

You have been warned now. Are you ready to consider that taste may not be personal? That the concept of basic tastes is ready for the dustbin? That gastronomy is even more than (as the *Oxford Dictionary* states) "the practice or art of choosing, cooking, and eating good food"? That there is no fundamental difference between white and red wines? That there may be a comprehensive theory about the combination of any drink with foods? That this same theory also gives guidelines about the order of dishes in a menu and even on the composition of the dishes itself? If you feel you are ready, then there is one last thing I ask you to do: question everything you know about gastronomy—or better still—forget everything you know and start anew. Let this book be your compass for finding your way into the gastronomic world.

Acknowledgments

This work could not have been completed without the support of others. I must first mention my family, who have always granted me the time to do my work, whilst following me and giving me their attention at the same time. They have never objected to my gastronomic experiments. From a family perspective, my father deserves a special mention. He did the original groundwork in the initial wine research, infecting it with his open mind and his enormous practical knowledge. He also gave me the freedom to find my own way. Neither of us ever suspected that it would culminate in this work. Although my father has passed away, those who knew him have no problem imagining how proud and grateful he would be.

The important role of my students should also be mentioned. In the early days, they were essentially my colleagues—among them renowned chefs and sommeliers. Why should I be able to teach them anything? Their skepticism did in fact force me to delve deeper and inspired me to find the answers to the questions they had. Later, students helped me with clarification and gave me valuable feedback on the text, making it more accessible and easier to understand.

With regard to content, four men and one woman have made essential contributions in different stages of the development of the theory and in various aspects of this volume.

Bob Cramwinckel was there at the beginning. He initiated the wine research and remained actively involved until the work was completed. He also gave intellectual inspiration whenever it was needed.

Professor Wim Saris placed his trust in me when he became my doctoral advisor, and he critically guided me throughout the process.

Jan Gispen was hugely supportive in establishing the Academy for Gastronomy, in helping set up the courses, and in finding the early answers.

Jan Schulp is a modern-day *Homo universalis*. With his profound and versatile knowledge, he made important contributions to this text.

The one woman who was an essential contributor is my wife, Carla. Her intelligence and unstoppable quest for logic has forced me to consider and reconsider, to ponder, and to dismiss ideas and accept others.

I am indebted to all of these individuals for their continuous support. As an expression of my gratitude, this book is dedicated to them.

PROLOGUE

In early science fiction novels, eating was reduced to not much more than taking a pill. It contained all the nutrients and energy we needed and was taken within a second. This science is no longer fiction. Astronauts survive on them. And on Earth, the labels of our processed and pre-packaged foods show long lists of ingredients, most of which have names that a normal person can hardly pronounce. We are able to make a potato chip taste like roasted chicken with thyme. The nutritional values are cited as are the energy it contains in kJoules and kCalories. Therefore, we know exactly how many (milli)grams of fats, proteins, carbohydrates, vitamins, minerals, and so forth our food contains, and often the percentage of daily value is also indicated. These days, consumers are better informed about nutrition than ever before, and many people also realize that healthful nutrition is of extreme importance to our bodies. Our biological system needs to be maintained.

Clearly, there is nothing wrong with knowing what our food contains. It is just another example of the progression of the human race. But just as with other advances, there is a downside as well. We have shifted from food to nutrients. In his book, *In Defense of Food*, Michael Pollan introduced the term "nutritionism" to make us aware of the negative side. Basically, the term indicates the shift in focus from real foods to nutrients, which often implies a move from nature to industry—or if you prefer—from old-fashioned beliefs to sophisticated food science. We now tend to focus on nutrients that are supposed to be either good or bad for us. For millennia, we survived just on knowing the differences between foods and poisons. Simultaneously, with this shift in focus, the number of food-related health problems has vastly increased. Obesity is growing into a worldwide epidemic; cancers are abundant, as are food allergies. We may be glad that we know so much more about how to treat many of these diseases so that we actually get older at the same time.

Nevertheless, we seem so preoccupied with the content of our food that we tend to neglect the other important aspects of feeding ourselves: the pleasure of eating, its social function, even its taste. Our focus on food has turned from holistic and cultural to instrumental and technical. Pleasure, taste, and social interaction are concepts that do not fit in the instrumental approach. And yet, they do play a major part in our daily lives. Is our food intake not much more than keeping our biological system functioning? From the earliest of times and onward, dining, the combined enjoyment of food and drink, has been extremely important to human beings. The cultural, social, emotional, and hedonistic elements of food are indisputably valuable but never mentioned on the label and in the product descriptions. One may start to worry that our evolution has caused some people to think that these emotions are illusions. They do not realize that our taste sensations are directly connected to the primary part of our brains—the part that also rules our other emotions.

The emotional and social aspects of our daily bread have always played a role and cannot, nor should, suddenly be negated. The social part of nutrition is too valuable to lose. Just look at the origin of the word "company." It is derived from the Latin *con pane*: with bread; eating together! Let the science of food pills remain in the realm of fiction. If not, we would seriously shortchange ourselves.

Meals can be memorable . . .
Food can move us . . .
Wines can leave lasting impressions . . .
Tasting can give pure enjoyment . . .

Gastronomy can in some respects be considered as the antonym of nutritionism. It is a holistic concept. It focuses not solely on the food and its composition but on the human that eats it as well. And, in doing so, gastronomy is especially interested in why this human likes the things he eats or drinks. This requires that we know how we taste food, which is complicated because tasting is by definition subjective as it is the subject that tastes. Furthermore, we use all our senses to register taste (or rather flavor, as will be explained later), and there are quite a few factors that may influence this registration, many of which have hardly been studied. This makes the study of gastronomy quite challenging and interesting.

Understanding taste and flavor and how a human reacts to it is not merely a hedonistic exercise. Gastronomy relates to big societal issues such as nourishing the elderly and the food children eat at school—concerns that are certainly not trivial and hardly hedonistic. And in the hospitality industry, where people pay for their food and an accompanying beverage, it is highly functional to ascertain that people are going to enjoy what they bought. It is the basis of commercial success; but also outside of the traditional industry, gastronomy merits much more interest.

CHAPTER 1

Introduction

2 THE ESSENCE OF GASTRONOMY

THERE IS NO REASON TO
ARGUE ABOUT TASTE

There is a lot of appeal in being active in gastronomy. Anyone who loves foods and beverages will quickly agree to that. Yet, however interesting it is and always has been it is also a new scientific field that may even be one of the most ill-defined. The origin of the word is Greek and refers to the laws of the stomach; that surely does not sound like a feast, yet there is no reason to doubt that the ancient Greeks were good gastronomes.

Gastronomy is not merely about culinary enjoyment or "the practice or art of choosing, cooking, and eating good food" as the *Oxford Dictionary* states. A gastronome is also much more than a foodie. He or she can—and in relevant situations should—be a modern and broadly trained food professional who knows a lot about flavor, taste, and tasting. Indeed, a gastronome understands what happens to proteins while boiling an egg and realizes that the same processes take place in the preparation of meat or fish: in a few minutes, it changes from juicy to dry. He or she also knows that this change in texture

1

may influence the combination with a wine. This implies that there is need to explore the flavor of a wine (or any other drink for that matter) and all kinds of foods, to understand the flavor of both, in order to be able to find the right combination. By its nature, flavor can be assessed objectively, at least to a certain extent. Nevertheless, preferences differ from person to person wherever you are in the world, and they vary across cultures.

In this book, gastronomy is introduced as the science of flavor and tasting. The title of this section is "there is no reason to argue about taste." If we would add "anymore" to this statement, we have a good start for introducing this new science. For a long time, most people have considered taste to be a personal perception—an individual experience that everyone defines for him- or herself over and over again. Why elaborate on something as volatile and frivolous as that? Consequently, our sense of taste has not received nearly as much attention as our other senses, and in the twenty-first century there remains a lot to be researched. Fortunately! At the same time, it is remarkable that we have accepted this for so long. It is not even very logical. A steak tastes different than a baked salmon, and also coffee and tea are very different. Wine experts may surprise people by taking a sip and telling the origin and quality of a certain wine. Thus considered, taste is a product characteristic rather than a personal conception. The word "taste" as it is generally used is rather ambiguous. Therefore, we will first elaborate on the definition of the concepts involved. An old saying suggests that you should not compare apples and oranges, yet sometimes it helps if you do. In fact, we should have started doing that a long time ago.

STRUGGLING WITH WORDS

When you compare the words *color* and *taste* in the dictionary, the ambiguity of the word *taste* becomes evident. The *Oxford Dictionary* defines color as "the property possessed by an object of producing different sensations on the eye as a result of the way it reflects or emits light." Taste is defined as "the sensation of flavor perceived in the mouth and throat on contact with a substance." Why is taste a sensation and not a property of an object, such as color? The clarifying sentence that the dictionary uses is "the wine had a fruity taste." This suggests that the fruity taste is—such as color—indeed a property that the wine possesses that is registered through our mouth and throat (and nose, I would add). There is no fundamental reason to treat taste differently than any other sensorial experience. Just the same, sound is a property, and the sound waves are registered by our ears.

There is something else, however. Unlike with *color* the dictionary does not stop there. Taste is also defined as "the faculty of perceiving taste." This is where the ambiguity sets in. Taste perceives taste; isn't that weird? The dictionary exemplifies this by stating that birds do not have a highly developed sense of taste. With regard to color, we would speak about "sight." This refers to the sensorial capacity to see things. Apparently taste is both the property and the sensorial capacity. And, the indistinctness of the word gets worse. Taste is, as we all know, also used in regard to a person's personal liking. The dictionary illustrates this as "this pudding is too sweet for my taste," or "have your lost your taste for fancy restaurants?" So taste can be many things at the same time: a property, one of our senses, and a preference. If gastronomy is presented as the science of taste and tasting, the first challenge is to better define some of these notions involved and to see how these concepts are related. This is a crucial step.

A part of the solution is found in the initial definition. Look back at: "taste the sensation of flavor. . . ." We must give meaning to the word *flavor* and clearly distinguish it from *taste*. The *Oxford Dictionary* clarifies flavor as "the distinctive taste of a food or drink." The official sensorial definition (from the British Standards Institute) elucidates: "the combination of taste and odor, influenced by sensations of pain, heat, and cold and by tactile sensations."[1] This definition implies that the senses of taste, smell, and touch are involved in registering flavor. Or, turning it around, everything we eat and drink has gustatory (taste), olfactory (smell), and tactile (touch) components. Last mentioned can best be broadly defined: everything that is registered by our sense of touch, comprising all kinds of textures, pain, and temperature. We may conclude for now that flavor is the result of product characteristics and can be considered as a technical, objective concept. Our senses do the tasting and communicate the elements of flavor to the brain.

We propose to go even one step further. The aforementioned elements can be considered as the intrinsic components of flavor. But before we take a bite or a sip we already have seen or heard things about what we are going to taste, and we have our recollections and prior experiences—implying that there are also extrinsic elements. They may be more important than often thought. "What you see is what you get" finds a new dimension in flavor perception. We will see that if yogurt is colored red with neutral coloring, people tend to perceive that it is sweeter. A strong brand can also positively influence registration just as a bad recommendation may negatively influence our perception. If these extrinsic factors of flavor are taken into account, tasting gets even more multisensory—external clues mainly registered by sight and hearing are involved as well.

This is all the more reason to use the word *flavor* whenever an edible product is involved. It is the flavor, and not the taste, of a certain dish or a wine that is being tasted. Coffee, tea, tomatoes, and chocolate have flavor, just as a banana and spareribs do. Flavor is a product property. At the same time, flavor may evoke emotions, which leads to preferences and appreciation. To mention just one example: in some cultures, insects are a delicacy; in others, people would not dream of eating them. Tasting, rather than taste, is the process that registers flavor. Tasting is a sensorial activity just as hearing, smelling, touching, and seeing. And indeed, it is possible that the flavor of the wine is not to your taste.

SOURCE OF PLEASURE

Although there is a strong biological drive to eat and drink, in order to keep the human biological system going, it is the flavor of these "intakes" that provides the extra dimension. Flavor gives meaning and pleasure to them. The rate of pleasure may differ from person to person, from situation to situation, and from culture to culture, but the function of flavor in relation to food is the same: it is the discriminating factor, the basis of food choice, preference, and intake. Consequently, everybody has experience with flavor and tasting, with personal likes and dislikes, and with preferences for certain foods and drinks in specific situations. People do not eat or drink just to survive, and it may safely be assumed that tasting plays a predominant role in everybody's daily life. The "right" foods and beverages provide an important source of pleasure in that daily life. In her report on the development of food preferences, Birch[2] concluded that the only activities of daily life rated higher than "a fine meal" were (a) spending time with family, (b) holidays, and (c) sex. Evidently, the order may vary from person to person.

Gastronomy has a lot to do with the art of living, enjoying the beautiful things of life. For many people, wines and good food are an essential part. The French connoisseur, Jean-Anthelme Brillat-Savarin, was the first one to publish a book on gastronomy: the *Physiologie du Goût* (Physiology of Taste; 1825). Whoever is involved with gastronomy, either privately or professionally, will use the name Brillat-Savarin with due admiration. The "Aphorisms of the Great Teacher" are very well known. ("Great Teacher" being the way in which, tongue-in-cheek, Brillat-Savarin refers to himself.) He writes, for instance: "The fate of nations depends on the way they eat"; "tell me what you eat: I will tell you what you are"; and "the discovery of a new dish does more to the world than the discovery of a new star." And also: "Animals feed; people eat, only the man of intellect knows how to eat" and "Drunkards and guzzlers do not know how to eat or drink."

It is remarkable that this book, being over 150 years old, still is considered to be one of the most important books written in the world of gastronomy. Have there been no developments in this field, no special discoveries, no research, ever since? Yes, incredibly fascinating developments have taken place, in the kitchen as well as in the world of wine. Suppose it would be possible to resurrect Brillat-Savarin and take him to one of the famous restaurants of today. We would introduce him to the work of the winemakers and great chefs of our time. Undoubtedly, he would start comparing past and present and, given the chance, would write a new gastronomic standard work because many of his intuitive suggestions can now be based on solid science. To his great surprise, he would learn that it took 149 years before gastronomy had its own professor. At the time of his publication, Brillat-Savarin suggested this would not take more than 30 or 40 years.

TIMES HAVE CHANGED

The "Great Teacher" would agree that most of his views are ready for revision. And also that the gastronomic rules that people use these days are rather silly. You know the rules: white wine goes with fish; red wine goes with meat; drink white before red, young before old. These are phony promises. You realize that when you consider what you are saying. Take "white wine with fish." If you really believe that, you are saying that all white wines are the same. Otherwise, the equation could never stand. Clearly, there are many differences in white wines. For those who need time to consider, start at the other end. Are all fish created equal? Is salmon like turbot, trout, or eel? And then how is your fish prepared? Poaching, baking, or deep-frying it will lead to different flavor profiles; sauces and garnishes may change the profile even more. In practice, it turns out that certain white wines are delicious with red meat just as red wines may be very tasty with fish. The rule to drink white wines before reds falsely suggests that these are lighter than red wines. Not the color, nor its age, but the flavor richness of the wine determines the order and if meats are to be served before fish in a menu. These rules have not lost their meaning because the world has changed. These were always empty shells—and please note that they were not suggested by Brillat-Savarin. One of the few recommendations that he made still stands—have lighter wines before richer ones.

There is a complicating factor. In Brillat-Savarin's times and up to about the 1970s, the preparation of dishes was likely to be rather standardized. Cookbooks of the time, and especially the *Guide Culinaire* from Auguste Escoffier, told you how to do things, and every dish or preparation had its own (often French) name, such as the tournedos "Rossini," canard à l'orange, sole Véronique, pêche Melba, and poire "Belle Hélène." If you ordered it in a restaurant, you knew what to expect. The quality of the chef could be judged by the way the dish was prepared—the closer to the original, the better. These days we expect the best chefs to come up with their own versions and combinations. We expect personality rather than perfect imitations. Just as many other things, flavor has been emancipated and is liberated from its old conventions. And not only in food. In

wines, a similar development took place. Modern winemakers are also freethinkers. Their orientation is the world. They compare their wine to other references from all over the world, just as chefs do in the kitchen. New dishes, inventions, and experiences spread over the world within days through Twitter, YouTube, and the Internet at large. To succeed in the flavor business, you sometimes have to dare to do things differently, to use modern techniques, and, in some cases, to defy regulation, as may be the case in appellations. As a result, in the world of wine and food, more has changed during the last 30 years than in the 2,000 years before. Consequently, the labels of many modern wines have become poor descriptors of the flavor to be expected, and the same can be said of the descriptions of dishes on the menu. Basically, you cannot know what you are going to get without prior experience.

Something else has dramatically influenced the things we eat or drink: the availability of ingredients. Fast transport by air and greater cooling capacity enable us to enjoy practically anything throughout the year. This has run out of hand so that now we are back to promoting local foods and customs and to putting food miles on the label to make people at least realize what needed to be done to get a certain fresh ingredient on the shelf. Fish from Chile, asparagus from Tanzania, game from New Zealand, strawberries from Spain, tomatoes from a greenhouse wherever in the world—it is all freshly available. If you want it, you can have it, no matter the part of the world or the season you are in. Throughout the year there will be a climatologic zone somewhere in the world that can provide the product you might be looking for. This book is not about judging the pros and cons of this situation. It asks you to realize that in these days we are able to make choices that we could not make before. And that—consequently—we no longer can count on our experience and intuition based on the past.

This abundance of choice is making everybody's gastronomic life harder rather than easier. If there is not much to choose, you eat what there is. In food and wine, this has been the case for ages. It was a regional and seasonal affair. The suggestion that regional wines will always go well with the regional food finds its origin there. The truth is that in daily gastronomic life such combinations may very well be disappointing.

It all leads to one big conclusion: understanding flavor in all respects and the knowledge about ingredients are more important than ever before. All suggestions that are based on the past are ready to be scrutinized. It is time for a new and embracing theory that fits our time and gives answers to the gastronomic questions of today and, preferably, also those of tomorrow.

SETTING THE STAGE

"It is inconceivable that gastronomy, before too many years, will not have its own academicians, its professors, its yearly courses, and its contests for scholarships. First of all, a rich and zealous enthusiast must organize in his own home a series of periodical gatherings, where the best-trained theoreticians will meet with the finest practitioners, to discuss and penetrate the branches of alimentary science."[3]

This is what Brillat-Savarin predicted in his book published in 1825. It took quite a bit more time than he thought. Nevertheless, in the last couple of decades, it happened after all. One of the first books that elucidated "the branches of alimentary science" was Harold McGee's *On Food and Cooking: The Science and Lore of the Kitchen*. It was first published in 1984 and revised in 2004. I remember reading it like a novel; it kindled my interest in food science during the time when I had just started as a restaurateur. The epic work of Nathan Myhrvold and his team, *Modernist Cuisine: The Art and Science of Cooking*, appeared in 2011 and will undoubtedly be an inspiration for the generations of curious cooks to come.

I published my first (rudimentary, if you look at it now) book on gastronomy in 1998. My thesis and consumer edition date from 2004. Simultaneously, but unknowingly, others started developing building stones for a new approach. A neuroscientist, Charles Zuker, challenged the assumption that taste cells were "broadly tuned," meaning that cells carry sensors for all "basic tastes." He supposed instead that we would have separate receptors for basic tastes and went looking for them.[4] In 2000, a whole range of bitter receptors were identified,[5] and in 2001, the sweet receptors.[6] It triggered a whole new way of looking at cells and receptors and led to finding "new" basic tastes such as fat, calcium, carbonic acid, and metallic. Some of their receptors still have to be established, and there are likely more to come. Another neuroscientist, Gordon Shepherd, follows the neural signals to the brain. His book, *Neurogastronomy: How the Brain Creates Flavor and Why It Matters*, was published in 2012. Note that he uses the words *flavor* and *gastronomy* in his book title. It is important to understand how our brain functions and how all the sensory signals that it receives when we eat or drink are bundled together to "create" the perception of flavor.

Progress was made not only in neurology. Carolyn Korsmeyer published her well-documented philosophical work, *Making Sense of Taste*, in 1999. She convincingly argues that tasting merits more respect and attention and explains why—up to now—it had to do without. In the meantime, great chefs discovered how science can help them make new and spectacular dishes. Ferran Adrià (restaurant: El Bulli, Rosas, Spain), Heston Blumenthal (restaurant: The Fat Duck, Bray, UK), and René Redzepi (restaurant: Noma, Copenhagen, Denmark), among others, are modern chefs who popularized new techniques that also brought them to the center stage of world gastronomy. Scientists such as Hervé This (Collège de France, Paris) have done their share in explaining what really happens in cooking. *The Kitchen as Laboratory: Reflections on the Science of Food and Cooking*, edited by César Vega, Jos Ubbink, and Erik van der Linden (2012), epitomizes the state of the art in what is called "molecular gastronomy." At Wageningen University in the Netherlands, a new master study—advanced molecular gastronomy—was introduced, and at Stenden University in Leeuwarden (also in the Netherlands), I was installed as professor of gastronomy in 2011.

Within just a decade, flavor and tasting developed from being figurants in the world of science to actors that attract attention. When this all started, it was inconceivable that the esteemed journal *Nature* would publicize a supplement on flavor, yet it did so in 2012. That same year, Peter Barham and Per Møller published a dedicated online scientific magazine, *Flavour*,[7] to share new findings. Gastronomy, the science of flavor and tasting, is on a roll.

GASTRONOMY, THE SCIENCE OF FLAVOR AND TASTING

Science is defined by the *Oxford Dictionary* as "the intellectual and practical activity encompassing the systematic study of the structure and behavior of the physical and natural world through observation and experiment." What are the ingredients of gastronomy as a science? Where are the building blocks to be found? We have seen that many aspects play a role; some of the concepts are intertwined. Therefore, we need to look closely at the concepts that are involved and at their meanings. Figure 1.1 visualizes the interrelation of flavor, tasting, and perception. The perception is the image the brain creates using the signals of all the senses concerned: mainly gustation, olfaction, feeling, and also hearing and vision. It is closely related to appreciation, either positive (liking) or negative (disgust).

The flavor of a food or a drink has two sides: a technical and an emotional one. Product specialists and/or sensory panels can objectively assess the technical aspects. It is imperative to pay attention to the use of statistical techniques and the interpretation of results. Furthermore, there are instruments available, such as the electrical nose and electrical tongue, which are getting more and more sophisticated; it is possible to measure the texture of fluids by rheometry and all kinds of aroma components with a gas chromatograph. These are examples of instruments that can do the tasting or at least a part of it. Ultimately, they can replace human beings as far as the objective side of tasting is concerned. However, the consumer assessment of a particular flavor may

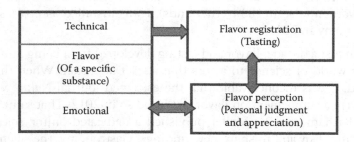

FIGURE 1.1   THE INTERRELATION OF FLAVOR, TASTING, AND PERCEPTION.

not be the same. Tasting refers to the reception of the sensory message; how this message is interpreted is quite another matter. Nevertheless, tasting has been the focus of the bulk of the academic interest.

Consumers are not neutral observers and, therefore, flavor perception merits our attention. Every human being has a personal framework in which he or she tastes. There are differences between men and women, children and the elderly, and genetic variations of human beings in general. Culture and experience play a role as do the climate and the price of products, just to mention a few of the many influences. And it does not stop there. Products that are tasted simultaneously tend to react on each other. There are tastes that enhance others and some others that degrade. Even the greatest of wines can turn into an ordinary or even bad-tasting wine with the wrong dish, and adding an unfitting herb or spice may spoil the flavor of a dish. Surely, those types of interactions can better be avoided.

Next to the individual, the flavor has to fit time and place. Consequently, the relationship between product quality as such and consumer appreciation is indirect and can be very complex. It is conceivable that even though product quality may be high, based on product characteristics, the consumer will not like it. In other words: a product is not good, but it is found to be good. This concept is known in marketing as the difference between the product approach and the consumer approach, and it applies to many other product categories as well.

To better understand the liking or disliking of a certain flavor, it is essential to know more about the relations between flavor itself and how it is registered and perceived. In Figure 1.1, the arrow at the emotional level between flavor and liking goes in both directions to indicate that external factors such as a brand, a package, a color, or a shape can influence flavor registration. A strong, well-known brand is likely to have a positive impact on flavor perception just as the name and fame of a chef influences the flavor perception of a dish. In a negative way, foods or drinks that supposedly made you sick one day are not liked and even avoided. This assessment takes place in the human brain and is based on the sensorial signals that are gathered there.

From the sensorial perspective, flavor registration is quite different compared to seeing, smelling, and hearing. Those are simple senses, whereas

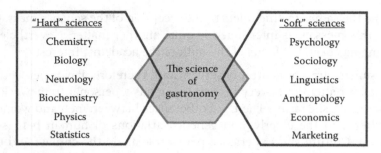

FIGURE 1.2   GASTRONOMY: INTERDISCIPLINARY SCIENCE.

tasting is a multisensory experience and therefore arguably the most intriguing and certainly the most complex of our sensorial experiences. The sensorial registration of flavor relies upon cooperation of the senses and, in the brain, upon a synthesis of sensorial inputs. We taste with our eyes, nose, ears, and yes, also with our tongue, or rather with our mouth. Partly because of this multisensory character, tasting is a complex and difficult matter to investigate. We will elaborate on that later.

For now, the conclusion may be drawn that gastronomy demands a versatile approach. In many fields of science, valuable contributions are found that elucidate elements of flavor, tasting, and appreciation. Gastronomy is a true liaison between the hard sciences (such as physics, chemistry, and even food science) and the soft sciences (such as psychology, sociology, and marketing). The scientific challenge for gastronomy is to gather as many pieces of useful knowledge and to find how they all fit together. This truly interdisciplinary approach characterizes the science of gastronomy (see Figure 1.2).

CHEMICAL SENSES AND CLASSIFICATION

Smell and taste are considered the chemical senses. They are closely related in more than one way. First, smell is an inseparable part of taste; a lot of what is called taste is in fact registered through the nose. While chewing and eating the food, odorants are liberated in the mouth that trigger the so-called olfactory system. Second, people tend to undervalue both senses. In ranking the importance of their senses, people put sight and hearing above smell and taste. Illustratively, the words *anosmic* (total or partial loss of the sense of smell) and *ageusic* (no perception of taste) do not belong to the everyday vocabulary, such as *blind* or *deaf*. It can even be argued that people with poor tasting or smelling abilities might not be aware of their problem. Loss of sight or hearing is quickly apparent to the person and to others. In general, a loss of smell or tasting capacity occurs gradually and is often undetected. A diminished interest in food can result, leading to weight loss and, ultimately, undernourishment. Methods for treating such chemosensory losses are greatly needed.[8]

In general, smell and taste seem to lag behind the other senses; less is known, particularly in regard to taste. The complexity of the multimodal registration of flavor, with its interactions of all kinds of sensory signals, may not be the only reason. The chemical senses are generally considered to be inferior to vision and hearing. Perhaps they have always been too closely related to the pleasure side of life. Hedonism or epicurism are not considered very serious philosophies. Now that the attention for food and its nutrients has gained popularity, including aspects such as how they affect our health and why we like certain foods, there is every reason to know much more about the functioning of the related senses.

The chemical senses are hard to classify. Colors and sounds can be described by their objective physical qualities, such as wavelength or vibrations per second (millimicrons, Hertz) and intensity (such as lux or decibel). The recognition and subsequent definition of these physical properties showed their structure and made them measurable. In other words, it revealed their objective side, which sets them apart from what people think of them. In other words, you may like a certain color or not, and you may listen to the music you like, despite their objective qualities.

To be able to discuss the objective side of flavor, we need measurable concepts such as frequency and intensity. For a long time, these have been lacking. The focus has always been on the so-called primary or basic tastes. Looking at any basic biology book, you are led to believe that taste can be reduced to sweet, acidic, salty, and bitter. For centuries, research has been conducted on substances, and yet it has not led to a solid and fundamental understanding. The reason is that the hypothesis of the basic tastes is rather flawed.

The suggestion dates back to the time of Aristotle. In addition to the basic four, he also listed spicy, astringent, and harsh/rough as basic flavors. Strangely enough, the last three mentioned have not survived in history. Indeed, they are quite different and related to what we now call "mouthfeel." In the sixteenth century, the French physician Jean François Fernel rightly suggested that taste buds are sensitive to fat and also suggested to add "tasteless" as a basic taste. Later still, alkaline and metallic were mentioned as basic tastes, but all suggestions were discarded; the focus stayed on the basic four. This changed in 1908, when the Japanese chemist Kikunae Ikeda discovered umami, or the taste of glutamate. It is now referred to as the fifth taste. Something else is remarkable. In Germany, David Hänig reported in 1901 that there are no zones on the tongue. Nevertheless, these continue to be mentioned in all kinds of textbooks. On our tongue, different structures of papillae can be distinguished, but specific zones where certain flavors supposedly are tasted are nonexistent.[9]

We had to wait until recently for a shift in thinking. Sophisticated techniques have enabled new research. With the use of functional magnetic resonance imaging (fMRI) nerve signals can be followed and linked to parts of the brain. We have learned the receptors on the tongue and in the mouth are much more capable than had been thought. Receptors have been identified for carbonation (CO_2), fatty acids, metal, and calcium, and there are likely more to come. With all these new basic tastes, the focus on the basic four becomes even more debatable. This can hardly be considered as a surprise. All our senses are involved in tasting. How could the singular focus on the papillae of the tongue ever provide a solid basis for taste or tasting? Other important factors need to be taken into account as well.

UNIVERSAL FLAVOR FACTORS

The modern view on gastronomy is scientifically validated in my thesis, "The Concept of Flavor Styles to Classify Flavors."[10] The parameters that made this all possible are mouthfeel and flavor richness. They will be elaborated upon in this book and are briefly introduced here.

MOUTHFEEL

Nowadays, it is hardly conceivable that the concept of mouthfeel was not known and was hardly used. Within mouthfeel, we distinguish the first two dimensions of flavor: contracting and coating. Acids and salts cause cells in the mouth to contract. The drying of certain bitters or absorption of saliva by starch and the crispness of crusts are also examples of contracting.

Mouthfeel coating is quite different. Here, substances leave a thin layer in the mouth. Think of fat and sugar in a solution. Also, the egg yolk of a soft-boiled egg is an unmistakable example of coating. Foods and wines are always intricate compositions of contracting and coating influences. Their flavor is the result of choices that chefs and winemakers have made. In some cases, the balance is more toward the contracting side; in other cases, it is more toward coating.

FLAVOR RICHNESS

The third dimension in our model is flavor richness, or flavor intensity. In the metaphor with sound, we would call it decibel. The analogy goes even a step further. Decibel is in indication of the volume (the quantity of energy per square meter) of sound but says nothing about the sound itself. Likewise, flavor richness must not be confused with quality. In other words, if the flavor richness is high, this does not necessarily imply that the quality is high. Flavor noise exists. It is also good to note that measuring flavor richness is much more complex than volume. We cannot measure it by simple counting the micromoles per liter.

Within the flavor richness we can also say something about the type of flavor. We distinguish fresh and ripe flavor tones. Apple and citrus fruits are nice examples of fresh, just as parsley or mint, or cucumber and fennel. Ripe flavor tones are those of ripe melon or pear, or of rosemary and garlic. Also vanilla, cinnamon, and clove are examples of ripe flavor tones. The word *ripe* as a flavor tone should not be confounded with the maturity of the fruit. A ripe apple or lemon is always fresh.

Preparation can have a big influence on the flavor type. Onions illustrate this nicely. A raw onion is clearly contracting and fresh. After some time in the oven, both the mouthfeel and flavor richness change. Stewed onions develop sweetness, and therefore they are more coating. In flavor richness the ripe flavor tones are predominant. There are many examples showing that ripe flavor tones increase with the rise of flavor richness as a result of cooking techniques. In the final model, flavor type is not specifically mentioned. This aspect of flavor helps to provide better insight into flavor richness and ultimately into the flavor style.

Mouthfeel and flavor richness enable the description of flavor in a rather objective way. That provides a wealth of possibilities. It is the basis of a model that opens, to a certain extent, the doors of the world of science! Science endeavors to build and organize knowledge in the form of testable explanations and predictions about the world. We have just objectively described a part of reality. When put in mathematical language, these newly formulated descriptors lead to the flavor styles model that can be used to classify flavors. The challenge of a model is to test it—again and again—and to try to improve it wherever possible. That too is science: the exchange of information in an effort to get an even better understanding of the reality.

GASTRONOMY AND HOTEL SCHOOLS

Above all else, it should be expected that hotel school students are able to ascertain that their guests enjoy what is being served. After all, these guests do the tasting and must pay for the service. Clearly, it is vital for a successful operation that guests like what they have consumed. However important, without specific knowledge, liking is an erratic concept. It does not only depend on the actual products and their flavor. People travel around the world more and more often. They experience other foods and cultures, and they learn about other cultures and eating habits through the Internet. Most likely they are influenced by these experiences and—whether we like it or not—they are inevitably more critical about the services rendered wherever they are.

This requires quite a lot from modern hospitality professionals. They must be well trained to be successful in their trade. There is a lot to learn in many areas. They should know about the influences of varieties of edible plants and breeds of animals and agricultural methods on flavor. They should know something of the physical and chemical sides of products. They should be aware of the influences of preparation techniques on taste—and not only in the kitchen. The way all sorts of drinks are made also belongs to the field of study. This includes the mutual influences of products consumed at the same time. Above all, they must understand the client—an understanding in all respects, cultural, psychological, sociological, and so forth. The professional gastronome is a well-trained professional who can lead organizations or parts of them where foods and drinks are served. The gastronome is neither the chef nor the sommelier, but he or she understands their language and can communicate with them. Nor is a gastronome a food scientist, yet he or she knows enough of the processes involved to ensure that something tasty appears on a plate or in a glass. He or she is an all-around professional.

STRUCTURE OF THIS BOOK

Many of the subjects mentioned in this introduction will be elaborated upon in this book. Chapter 2 is all about flavor perception, the sensorial act of tasting, how it works, and what neural systems are involved. Chapter 3 will focus on understanding flavor with an in-depth discussion of the universal flavor factors and the new flavor theory. In Chapters 4 and 5, respectively, food and beverages are examined from the flavor perspective. What are the effects of ingredients and the techniques that are used? In the next two chapters, liking is examined. In Chapter 6, this is primarily done at the flavor level. For instance, practical guidelines are given for matching food and beverages. In Chapter 7, we will look at the interpretation of sensorial signals in the brain and discuss issues such as food choice, preferences, liking, and palatability. Figure 1.3 illustrates this approach.

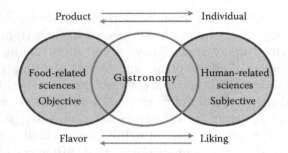

FIGURE 1.3 THE PRODUCT AND HUMAN SIDE OF GASTRONOMY.

SUMMARY

Gastronomy is the science of flavor and tasting. The word *flavor* is used, rather than taste, to make a clear distinction between the two concepts. In this book, the word *flavor* is used as a product quality, and it is interpreted in the widest sense, meaning that all elements that contribute to a flavor are included. Tasting is the human registration of flavor. All senses are involved in tasting, which makes tasting rather complex to understand. Furthermore, it is by definition subjective. Flavor, on the other hand, can be assessed objectively. It can also be classified with the parameters mouthfeel and flavor richness.

Products have flavor and people taste them. Gastronomy as a science requires an understanding of products and people with regard to eating and drinking. As such, it constitutes a liaison between natural sciences such as (food) chemistry, physics, biochemistry, and so on and human-related sciences such as psychology, sociology, marketing, and so on. A gastronome is a well-educated food professional with a solid understanding of the processes that are involved in eating and drinking. The future gastronome is a potential leader in the professional world of food. He or she is able to guide the creation and development of innovative food and beverage concepts. His or her expertise will help others to make the right decisions that lead to a better tasting world.

NOTES

1. ISO (International Organization for Standardization) (1992). *Standard 5492: Terms related to sensory analysis*. Geneva, Switzerland: ISO.
2. Birch, L. L. (1999). *Development of food preferences*. Ann. Rev. Nutrition 19:41–62.
3. Brillat-Savarin, J. A. (1826). *Physiologie du goût*. Paris: Sautelet. American edition: Fisher, M. F. K., trans., *The physiology of taste* (1949). New York: Knopf.
4. Tsunoda, S., Sierralta, J., Zuker, C. S. (1998). *Specificity in signalling pathways: assembly into multimolecular signaling complexes*. Curr Opin Genet Dev. 8:419–22.
5. Chandrashekar, J., et al. (2000). *T2Rs Function as bitter taste receptors*. Cell 100:703–11.
6. Nelson, G., et al. (2001). *Mammalian sweet taste receptors*. Cell 106:381–90.
7. www.flavourjournal.com
8. Finkelstein, J. A., Schiffman, S. S. (1999). *Workshop on taste and smell in the elderly: an overview*. Physiology & Behavior, 66:173–76.
9. Bartoshuk, L. M. (1978). *History of taste research*. In Carterette, E. C. and Friedman, M. P. *Handbook of perception*, Volume VIA, Tasting and Smelling (pp. 3–18). New York: Academic Press.
10. Klosse, P. R. (2004). *The concept of flavor styles to classify flavors* (thesis). Maastricht University, the Netherlands.

CHAPTER 2

Making Sense of Taste

Before reading this chapter, consider this: suppose this was a fundamental book on color; what would you expect? Most likely information on light and the spectrum. It would not be strange to discuss the classification in primary and secondary colors, matching colors, intensity, and absorption. It could also contain a chapter on pigments, types of paint, and painting techniques. A chapter on emotional associations with colors would be fascinating (why is white for the Chinese the color of mourning, although in the West brides marry in white?). But honestly, would you expect a chapter on the way the eye functions and how color is interpreted in the brain? I bet you would not! Yet, here is a chapter on the neural pathways through which our brain is informed about flavor. And you are likely not to think twice about it. In flavor, it is quite common to discuss the tongue. In fact, in trying to get a better understanding, scientists have been searching for flavor in the neural system for a long time—as if the spectrum can be found by turning the eye or the brain inside out. Neither color nor flavor is found within us. Nonetheless, certainly in respect to flavor, it is very interesting to see how our neural system works and

21

how our senses cooperate to grasp flavor in all its facets. It may even be helpful for developing a better understanding.

TASTING: AN INTRICATE PROCESS

Tasting is the sensorial capacity to perceive flavor. This chapter focuses on the underlying neural mechanisms. How do we taste and what is involved in tasting? The International Organization for Standardization (ISO) definition of flavor gives an important clue with regard to tasting. It states that flavor is "the complex combination of the olfactory, gustatory, and trigeminal sensations perceived during tasting. Flavor may be influenced by tactile, thermal, painful, and/or kinesthetic effects."[1] This is the scientific definition and uses words that may not belong in the common vocabulary. Therefore, *olfactory* implies the involvement of the nose, *gustatory* is about the participation of the tongue (and mouth), and *trigeminal* refers to the nerve that mediates the sense of touch in the region of the mouth. Apparently the last mentioned aspects are found to be so important that the extent of these effects is elaborated upon (tactile, thermal, painful, and/or kinesthetic effects). The gustatory, olfactory, and trigeminal elements of flavor refer to three different sensorial systems, and therefore, the ISO definition implies that these three senses are directly involved in tasting: smell, gustation, and touch. The word *gustation* (from the Latin *gustare*) is used rather than taste to prevent misunderstandings. Gustation is "just" one of the senses that are involved in tasting.

As obvious as all of this may seem, there is one hurdle to overcome: the general opinion on the use of word *flavor*. In dictionaries as well as in many scientific articles, and despite the official definition, flavor is often considered to be only the combination of taste and smell. An opinion survey published in the *Journal of Sensory Studies*[2] confirms this. In general, sensory specialists agree that taste and smell are the most important contributors to flavor. The mouthfeel components of flavor are often not directly included and are considered as something extra. Nevertheless, these are very important and may even be at least as important as the other two together. Textural elements are a part of flavor, and therefore, it is essential to include touch, or rather mouthfeel, in tasting. Surprisingly, in his book *Neurogastronomy* (2012), Gordon

Shepherd[3] builds a strong case for an integrated view on the perception of flavor—including the function of the brain—but overlooks the role of mouthfeel. Thus, despite the common interpretation of the word *flavor*, we will stick to the ISO definition and include the trigeminal system. This chapter will make a start in showing just how important it is.

In tasting, foods and drinks trigger sensory nerve cells that send electrical signals to the brain. This is called transduction, which is defined as the process by which physical entities from the outside become represented as electrical activity in a sensory nerve cell. In transduction, the boundaries among the systems become blurred, making it practically impossible for a human being to identify which system was responsible for what part of the flavor registration. This is especially true for gustatory and olfactory elements. Nose and mouth function closely together and do so with amazing speed and seemingly little cognitive effort. Traditionally, the gustatory part of tasting has received intensive scientific attention. Many reports are available on the so-called basic tastes, most of which rely on detailed studies often involving all sorts of animal species. However, it cannot be expected that flavor will be fully elucidated with a sole focus on one of the sensorial systems. Flavor perception is the outcome of a true synthesis of sensory registration. In this chapter, we will elaborate on the neural pathways and remain largely at the registration level of papillae and other receptors of information. In Chapter 7, we move up into the brain, the electrical machine that processes all sensory information and informs and guides us.

TASTING: WHAT HAPPENS IN THE MOUTH?

When food is introduced into the mouth, the tongue positions and presses it against the palate. This gives textural information. The food is moistened by saliva, and a process of chewing and swallowing begins. This is the process of mastication, during which the food is fractured and deformed by a complex combination of compressing, shearing, and tensing. The tongue, cheeks, and lips play a major role by directing the food and making it ready to swallow. This process coincides with the release of tastants, odorants, and the assessment of all kinds of textural properties of the food. These concepts refer to the sensory system that registers the information.

Tastants are substances that are transduced by the taste cells of the gustatory system, odorants are registered by the olfactory system, and irritants and textural elements are processed by the trigeminal system.

Saliva plays an important role in this process. Secretion takes place in response to chewing and to certain flavor properties, such as pH. Saliva mixes with the food and acts as a lubricant. It also facilitates flavor perception. Furthermore, saliva contains alpha-amylase (to digest starch), lingual lipase (to digest lipids), and lysozyme, an antibacterial enzyme. Among subjects, there is a large variation in flow rate and composition of saliva. However important the role of saliva in tasting is, a general lack of knowledge of the role of saliva on perception is reported.[4] It is clear that this oral process is highly dynamic. Information is gathered and transmitted to our brain, where the final result is assessed and named: it is our perception of the real flavor.[5]

In the process of eating or drinking, a wide variety of chemical stimuli are released. These can stimulate sensory receptors. Some of these chemicals (or molecules) are volatile, others are not. The volatile ones can be odorants or irritants and are transported from the mouth to the nasal cavity. Volatile substances may be liberated when foods or drinks stay in the mouth for a while and in the process of chewing. The nonvolatiles can be tastants, irritants, or textural components. Some of these are hydrophilic and mix easily with saliva; others are hydrophobic and do not. They may leave a coating layer behind. These chemicals and molecules are registered by flavor receptive and irritant-sensitive areas in the mouth. Figure 2.1 illustrates this.

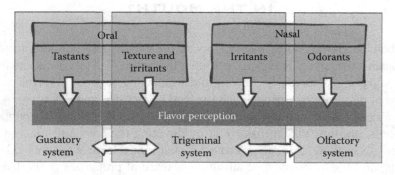

FIGURE 2.1   FLAVOR PERCEPTION AS THE SUM OF DIFFERENT NEURAL SYSTEMS.

People have different chewing behaviors, and it is not unlikely that these differences lead to different perceptions of what is tasted. It is reported that individuals with shorter chewing times tended to concentrate more on the most predominant properties of the food. Also they tended to perceive soft foods texturally as firmer and harder foods as less firm than individuals with prolonged chewing periods.[6] The next three sections will address each system at the receptor level.

THE GUSTATORY SYSTEM

PAPILLAE, TASTE BUDS, AND TASTE CELLS

Gustation and smell are considered to be the *chemical senses*, and the basic function of the gustatory system is therefore the registration of chemical components. Most named and well-known are sweet, salty, acidic, and bitter. More recently umami as "the fifth taste" has become better known. Registration is done by taste receptor cells (TRC), which lie within specialized structures called taste buds that are predominantly situated on the tongue and soft palate. Taste buds are onion-shaped structures containing between 50 and 100 taste cells, each of which has fingerlike projections called microvilli that poke through an opening at the top of the taste bud, called the taste pore. Chemicals from food, tastants, dissolve in saliva and contact the taste cells through this taste pore.

Most taste buds on the tongue are located within papillae, the visible small lumps that give the tongue its specific appearance. There are four different kinds of papillae:

- Fungiform (mushroom-like) papillae
- Circumvallate (wall-like) papillae
- Foliate (leaf-like) papillae
- Filiform (thread-like) papillae

Of these, the first three have taste buds. Last named, the filiform papillae coat the dorsal surface of the tongue and do not contain taste buds. They are responsible for keeping the food in place on the tongue and are involved in tactile registration.[7] The sensitive fungiform (mushroom-like) papillae are situated on the front part of the tongue. Generally, humans have about 200 of these papillae. They can be made visible after drinking some milk or placing a drop of food coloring on the tip of the tongue. Every fungiform papilla contains about three taste buds. The circumvallate papillae are larger and situated at the back of the tongue in the shape of an inverted V. There are only 12 of these papillae, but they each contain about 200 taste buds, representing about 50% of all taste buds on the tongue. The foliate papillae are situated in small trenches on the sides of the rear of the tongue. These papillae harbor about 600 taste buds on each side. Humans have about 6,000 taste buds, each with between 50 and 100 taste cells, totaling 300,000–600,000 cells. Besides in lingual papillae, taste buds are also found in the soft palate, uvula, epiglottis, pharynx, larynx, and esophagus[8] (Figure 2.2).

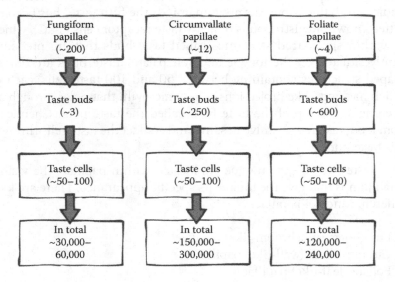

FIGURE 2.2   TWELVE CIRCUMVALLATE PAPILLAE ULTIMATELY REPRESENT 50% OF THE GUSTATORY FLAVOR SYSTEM. IN GENERAL HUMANS HAVE ABOUT 6,000 TASTE BUDS ON THE TONGUE WITH A TOTAL OF BETWEEN 300,000 AND 600,000 TASTE CELLS. THESE NUMBERS DIFFER FROM PERSON TO PERSON.

The taste buds of the fungiform and foliate papillae are connected to the chorda tympani nerve, which is a branch of the facial nerve. This nervous system is not only highly sensitive to chemical stimuli; it also responds to the cooling and warming of the tongue. The receptors of the trigeminal system are close to those of the gustatory system and therefore the signals get easily confounded. Salt and acidity are examples of gustatory flavors that are felt as well. Nevertheless the chemical transduction has received much more scientific attention than these textural aspects—or rather the elements that involve the sense of touch.

The taste buds of the circumvallate papillae on the back side of the tongue are connected to the glossopharyngeal nerve. This implies that gustatory information reaches the brain through two nerve systems. This is another explanation why people whose chorda tympani is damaged do not necessarily lose all gustatory information; there will still be transmission by the circumvallate papillae and the other systems. Figure 2.3 shows that both nerve systems cover about half of the gustatory cells.

PERSONAL DIFFERENCES

Taste bud cells are continuously renewed. The turnover rate of these cells is high; about every 10 days the cells have changed. This is highly functional because they are constantly getting damaged in the process of tasting.[9] Zinc appears to be an important element in keeping receptor cells functioning and in maintaining tasting abilities. Zinc deficiency is reported to lead to disturbances in flavor registration, caused by a decrease of the number of taste buds per papilla. People who want their taste system to stay in shape should see that they get enough zinc.[10] Cell regeneration slows down when people get older. Basically, this implies that older people generally lose sensory capacity; in hearing and vision, this loss is apparent.

The number of papillae and subsequent taste buds differs from person to person. In one study,[11] the number of gustatory papillae on the tip of the tongue averaged 24.5 papillae/cm^2, with a range from 2.4 to 80. Subsequently, the average taste bud (tb) density will differ as well. The

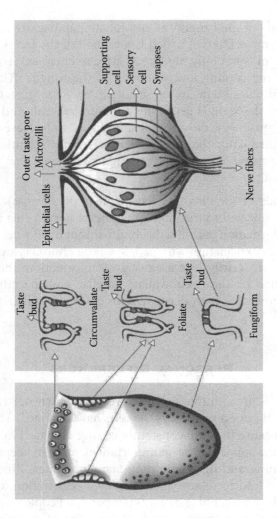

FIGURE 2.3   GUSTATORY "INSTRUMENTS."

fungiform-rich frontal region of the tongue attains adult size by eight to ten years of age. The posterior region, with the foliate and circumvallate papillae, continues to grow until 15 to 16 years. Densities and taste buds of fungiform papillae are reported to be significantly higher in eight- to nine-year-old children compared to adults. Furthermore, children's papillae are rounder and more homogeneous in shape.[12]

Consequently, it is likely that children register flavors differently, compared to older people. Parents ought to take that into account if a child does not like a certain food or drink.

In their study, Prutkin et al. reported that hormones in females also contribute to variation in flavor registration. Another human difference: African Americans reportedly show a tendency to prefer higher levels of sweetness in comparison to European Americans. Strategies aimed at influencing food preferences and dietary habits in individuals or groups should consider the role of these "genetic taste markers."[13,14]

As can be expected, differences in papillae and taste buds lead to variations in human gustatory sensitivity. In this respect, the so-called phenylthiocarbamide, or PTC, test is widely known in the sensory community. The number of fungiform papillae seems specifically related to the ability to discern the bitter-tasting chemical compounds PTC and 6-n-propylthiouracil (PROP). Some individuals do not taste their bitterness at all; others do but not as intensely as the group "super-tasters" that have the most fungiform papillae[15] (see Figure 2.4). In the general population, approximately 30% of individuals are non-tasters and 70% are tasters, some of whom are super-tasters. Alcoholism and smoking have been reported more frequently among non-tasters than tasters in some studies.[16]

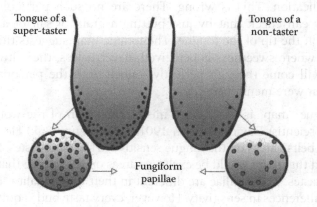

Tongue of a super-taster

Tongue of a non-taster

Fungiform papillae

FIGURE 2.4   SUPER-TASTERS HAVE MORE FUNGIFORM PAPILLAE.

Semantically, one may object to the terms *non-tasters, tasters,* and *super-tasters.* So-called non-tasters do taste; however, they do not register the substances PROP or PTC and may experience other substances differently. Reports are sometimes contradictory but seem to agree on super-tasters having a higher tactile sensitivity, which is in line with the observation that fungiform papillae are involved in trigeminal registration. Super-tasters tended to show less liking for highly sweet foods, just as they did for bitter foods.[17,18,19]

BASIC TASTES: PROBLEMS AND MISCONCEPTIONS

The gustatory system of flavor registration is responsible for the registration of the so-called basic, or primary, tastes. It has often been suggested that flavor is primarily about sweet, sour, salty, and bitter and that the papillae on the tongue are responsible for depicting these tastes—so far, so good, at least partly. The use of the term "basic tastes" is questionable because it suggests that these flavors are the building blocks that constitute flavor, just as the primary colors do. To prevent such misconceptions, it would be safer to address them as "gustatory flavors."

There are more serious matters with regard to the basic flavors. It is widely assumed that there are particular areas on the tongue that are specifically designed to register one of the specific basic tastes. Many textbooks feature the so-called tongue map and the where-do-we-taste-what indication. This is wrong. There are no such particular areas. Anybody can check that by just putting a grain of salt or a drop of vinegar on the tip of the tongue. The tongue map suggests that this is the area where sweetness is perceived; nevertheless, the saltiness and acidity will come through perfectly—albeit with the personal differences that were mentioned before.

The tongue "map" is based on a misinterpretation of the work of the German scientist David Hänig in 1901.[20] (See Figure 2.5.) He reported on "taste belts": areas on the tongue sensitive to the basic tastes. He never suggested that there would be exclusive areas on the tongue that register specific tastes. The papillae are different in their functionality. This may lead to differences in sensitivity. However, every taste bud is quite able to register all kinds of flavors—even more than the basic four.[21,22]

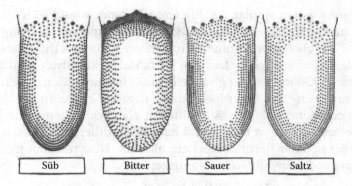

| Süb | Bitter | Sauer | Saltz |

FIGURE 2.5   THE ORIGINAL DRAWINGS OF TONGUE SENSITIVITY BY D. P. HÄNIG. (FROM HÄNIG, D. P., *PHILOSOPHISCHE STUDIEN,* 17, 576–623, 1901. WITH PERMISSION.)

Indeed, the focus has always been on the basic four, but as research methods get more and more sophisticated new basic tastes emerge. Japanese scientists identified umami, the flavor of glutamic acid, as a new, fifth basic flavor in 1908 (see the section "Umami" in Chapter 3). Glutamic acid is one of the 20 amino acids that make up the proteins in meat, fish, and some vegetables. It is also abundant in aged cheese and in other fermented products, such as soy sauce, oyster sauce, marmite, dry cured hams, and wines. Synthetic glutamate is used extensively in the food industry as a flavor enhancer in the form of the additive monosodium glutamate (MSG).[23] Other new flavors that have been identified and registered by papillae Include carbonic acid, fatty acids, and calcium. Neurons also respond to the oral texture of fat, sensed by the somatosensory (trigeminal) system. One should not be surprised if these findings and this new focus will result in the recognition of many other basic flavors, which will ultimately undermine this concept as a whole.[24,25]

The basic taste supposition is also undermined by another reason. Research on salty, acidic, sweet, and bitter is likely to be biased. In research, it is desirable to be able to compare results. Therefore, only a few specific chemical components have predominantly been used in sensory research: sodium chloride (NaCl) for salt, hydrochloric acid (HCl) for acidic, sucrose for sweet, and quinine for bitter. The choice for every one of these single substance clearly excludes others. Who says that results with a similar yet different substance would be the same? There are many different salts, acids, and bitter- and sweet-tasting substances. The choice for quinine to study bitter is perhaps the most questionable. Quinine is hardly found in foods and is extensively used in the pharmaceutical industry to give medicines a bad taste. Consequently, it is not surprising that bitter has gotten a negative image. In fact, bitter-tasting components have vastly different chemical origins. The bitter of chocolate, coffee, or beer would have given different results; it may even be a popular flavor component after all. New research methods have revealed many different bitter receptors.

And there is more. The so-called basic flavors are not even perceived as singular gustatory experiences; they are felt as well. Salt and acidity contract, while sweetness and umami have coating qualities. Some bitters are astringent and have a drying effect in the mouth. Neurologically and chemically, there is a poor relationship among the basic flavors. Substances that are chemically different may have a similar flavor. Interesting examples are to be found in sweet where, for example, aspartame, saccharine, and chloroform are perceived as being sweet although they have nothing in common chemically with sucrose. Furthermore, temperature and CO_2 have been reported to influence our perception of sweetness. The lack of common chemical structures suggests that there may be several different receptors and/or transduction processes.[26,27,28,29] In general, sweet, bitter, umami, calcium, and others react on the outside of gustatory receptor cells. Chemicals that produce salty and sour flavors act directly through ion channels, just as pain-like sensations register through nociceptors of the trigeminal system (see the section "The Versatility of Nociceptors" in this chapter).

In summary, after centuries of research, it is still not completely known how all the elements that constitute a flavor are transduced, and there is still quite some work in progress. To illustrate: the sour and salty receptors were not identified until 2006 and 2010, respectively.[30] Whatever progress is still to come, the concept of basic flavors is inadequate to describe flavor because it involves only the gustatory system. Focusing on this system alone (which has long been the case) without taking the olfactory system and the trigeminal system into account cannot explain the full picture.

THE OLFACTORY SYSTEM

THE OLFACTORY PROCESS

Olfaction can be divided in two aspects: the orthonasal and the retronasal smell. The first is the one we are best acquainted with. It comes from the outside and is registered by breathing through the nose. Retronasal smell refers to odors that are registered from within the oral cavity. Both "smells" are important for assessing the characteristics, richness, and complexity of a certain flavor. Nevertheless, the two parts of smelling have distinctly different functions. The orthonasal perception is used for identifying objects at a distance. It gives information about the potential palatability in positive (attractive smell) and negative (bad smell) situations. From an evolutionary perspective, it can safely be assumed that our nose (and eyes) have always guided us in making the proper food choices in times when there were no packaged, safe foods with elaborate labels. The retronasal perception (also referred to as the inner nose) contributes directly to flavor and, hence, food identification in the mouth.[31] (See Figure 2.6.)

The olfactory receptors are designed to capture odorants: small molecules that are light enough to be breathed into the nose and heavy and complex enough to be recognized. There are many volatile chemicals stimulating smell. Just imagine the diversity of the smells of citrus, cut grass, mint, various flowers, woods, fruits, herbs, spices, burnt aromas, sulfuric aromas, and ethereal, fatty, and sweet odors (such as vanilla). A certain "smell" usually consists of a complex mixture of many odorants. These are recognized individually by olfactory receptor proteins and assembled and identified in the brain.

Humans have approximately six million olfactory cells (about ten times the number of gustatory cells). Of the sensory systems, the number of cells is only exceeded by those of vision. Thus the qualitative range of olfaction seems quite wide compared to gustation. In general, women tend to have higher scores on tests of odor identification than men and retain their olfactory capabilities longer.[32]

In the process of eating, chewing, and drinking, structures break and change in temperature; odorants are released, and they may change

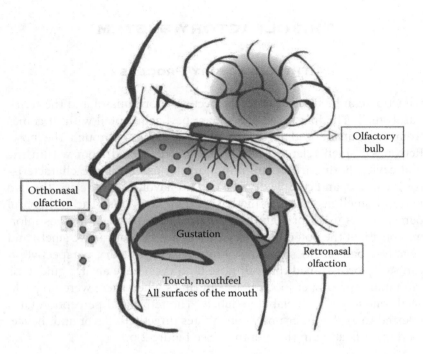

Olfactory bulb

Orthonasal olfaction

Gustation

Retronasal olfaction

Touch, mouthfeel
All surfaces of the mouth

FIGURE 2.6 THE TWO PASSAGES OF SMELL.

and also get lost. There is no debate as to whether olfaction has an important contribution to flavor registration, especially the retronasal part of it. Even the sense of touch is involved in the olfactory system. Trigeminal sensations such as burning, stinging, sharpness, and coolness can be produced by chemical activation of trigeminal fibers within the olfactory system.[33,34,35,36]

The importance of the role of smell in the perception of flavor is especially apparent when it is gone. When we suffer from a common cold, foods and drinks seem to have lost much of their flavor due to the fact that the nose does not function properly. This illustrates how closely the sense of smell is involved in tasting. In fact, smell and gustation lead to one unified perception and can hardly be distinguished from each other; they are intertwined. If the olfactory system is blocked, many products are identified differently, if at all. Figure 2.7 shows how products are perceived with (first bar) and without (second bar) the nose.

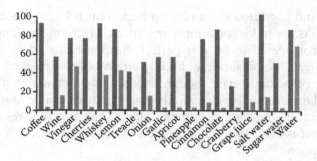

FIGURE 2.7   INFLUENCE OF THE OLFACTORY COMPONENT ON FLAVOR REGISTRATION. THE
SECOND BAR SHOWS PERCEPTION WHEN OLFACTION IS BLOCKED. (FROM VROON, P., VAN
AMERONGEN, A., DE VRIES, H., *VERBORGEN VERLEIDER: PSYCHOLOGIE VAN DE REUK*, AMBO
PUBLISHERS, BAARN, THE NETHERLANDS, 1994. WITH PERMISSION.)

ODOR COMPOSITION AND PREFERENCES

Ideally, the flavor follows the smell. The orthonasal smell gives a first
impression and a certain expectation. When a flavor is tasted, its aroma
should be rather similar. Appreciation is at stake when expectancies are
not met.

A natural fragrance that a subject can quickly identify with a single
sniff is usually composed of several hundred odorants. If fragrances are
produced synthetically, only the most important components (mostly
ranging from 10 to 20) are used. The hundreds of other odorants that
are present in the natural original add to the complexity of the real
fragrance. Instruments of analytical chemistry such as a gas chromatograph are increasingly sophisticated and good at qualifying the presence
and the quantity of fragrance components. Yet they do not perform as
well when it comes to identifying the key or dominant odorants of a
certain mixture. The human brain has the capacity to identify differences in a smell at a much more detailed level. In fact, millions of olfactory receptors are available to encode the components of a mixture. The
brain transforms this information into a unified sensory experience,
which is immediately associated with places and feelings and is also
compared to those previous experiences. It is one of the many examples
that illustrate the importance of distinguishing flavor from flavor registration and flavor perception, which was introduced in Chapter 1.

Context and cognition often play an important role. An oyster with a foul smell is not fit for consumption, although a strong-smelling cheese may be considered to be just perfect. Responses to odors can vary enormously. One person may find a certain odor unpleasant or even irritating or noxious even though another may not notice it or may even find it pleasing. However, only few references can be found about this aspect of smell, most probably due to the fact that research on this subject is complex because of personal, cognitive, and contextual interpretations.[37]

At the same time, some fragrances seem to be rather universally liked and are known to be appetizing, such as the odors of vanilla, roasting, and baking. The universal liking for these odors may even be innate, but little is known. It is also suggested that responses to basic tastes may have an unlearned component that is tuned to detection of ingredients (sugars are sweet, NaCl is salty, and many poisons are bitter). Olfaction is organized to identify foods more holistically, rather than to identify nutrients, and to be readily influenced by learning and experience.[38]

Although humans are very good at detecting odors, they generally have problems identifying and naming them. A possible explanation may be that the registration of odorants principally involves the right hemisphere of the human brain, while linguistic processing is predominantly under control of the left part.

OLFACTION, AGE, AND HEALTH

As human beings get older, progressive decline in olfactory ability has been reported. Above 60 years of age, people tend to have less of an ability to smell than younger persons. Between age 65 and 80 about half of the olfactory function is lost. People above 80 years of age usually have lost three-quarters of their ability.[39] Because olfaction is an important part of flavor registration, olfactory dysfunction is associated with decreased enjoyment of food. In the preparation of dishes for older people, this aspect could or even should be taken into account. However, the change in sensorial abilities may well coincide with changes in food preferences and beliefs about food (often in relation to health) as people get older. Therefore, the causal relation between the two developments is hard to establish.[40] A prerequisite is that the role of food in health and well-being is taken seriously. There is reason to doubt whether policymakers are aware of the importance of this role. Food is likely to be considered as a cost, not as an investment.

As soon as medication becomes a part of the daily routine, dramatic changes may occur. More than 250 drugs have been reported to affect flavor registration. There are many medical conditions or disorders associated with olfactory dysfunctions. Some examples are alcoholism and drug abuse, head trauma, HIV, anorexia nervosa (severe stage), and schizophrenia. Patients with very early symptoms of Alzheimer's disease experience a severe decline of olfactory abilities. Therefore, it is suggested that olfactory loss could be the first clinical sign of some neu-rodegenerative diseases. Memory-based olfactory tests may contribute to an early diagnosis of Alzheimer's disease.[41,42] It is safe to presume that such losses of olfactory capacity will also have a negative influence on the perception of flavor as a whole.

THE TRIGEMINAL SYSTEM

No Flavor without Some Kind of Mouthfeel

Next to gustation and olfaction, elements of flavor are registered, or rather felt, by the trigeminal nervous system. The trigeminal system is diverse, found all over the body, and refers to the sense of touch. In tasting, the trigeminal effects gather in the concept of mouthfeel, how-ever different the origin of the effect may be.

There are many everyday examples that show the importance and ver-satility of mouthfeel. These include the fizzy tingle of CO_2 in soda, beer, or champagne; the hot burn from different kinds of peppers; the tingling sting of ginger and some spices; the tenderness of a steak; the coolness of menthol; the pungency of mustard or horseradish; the bite from raw onions and garlic; the right creaminess or thickness of a sauce; the crispy dryness of toast or a crust of fresh bread; the soft juiciness of a ripe pear, melon, or peach; and the refreshing crispness of a ripe apple. In addition, temperature also contributes to mouthfeel.

Furthermore, it is widely accepted that fat is an important sensory textural element. There is evidence that groups of neurons respond to the texture of fat, while others respond to gustatory and olfactory inputs. It has now been established that the building blocks of lipids, fatty acids, are registered by the gustatory system.[43]

The "right" texture is closely related to quality. An apple is expected to be hard and crisp, and a pear, melon, or peach is best appreciated when soft and juicy. We do not like soft apples or hard pears. Sauces and soups need to have the right thickness or creaminess, and the palatability of many vegetables and carbohydrate-rich foods (such as pasta or potatoes) is also strongly related to texture. If toast or chips lose their crispness, they lose their appeal.

In fact, there is no flavor without some kind of mouthfeel. Apparently, our ancestors knew this very well. Etymologically the word *taste* stems from the medieval English word "tasten", which originates from the old French *taster*, meaning "to handle, touch, taste." It is probably derived from the Latin *tangere*, which means to touch. In contemporary Dutch, the word *tasten* still means to touch. Indeed, there is a lot to be felt in flavor: hard, soft, crispy, crunchy, effervescent, spicy, dry, viscous, fatty, puckering, warm, and cold are all sensations that are *felt* in the oral cavity. Even good old Aristotle thought of mouthfeel. He included spicy, astringent, and harsh/rough in "his" basic flavors. Nevertheless, this essential part of flavor has not received much attention. Mouthfeel may well be considered as the "forgotten attribute."[44] On the product level, mouthfeel is influenced by the texture of foods. The ISO definition of texture is "all the mechanical attributes of a food product perceptible by means of mechanical, tactile, and where appropriate, visual, and auditory receptors."[45] This all-encompassing definition shows that texture is a multifactorial quality and not a single entity. This section of food science is called colloid science.

THE RECEPTORS OF THE TRIGEMINAL SYSTEM

The list of mouthfeel examples shows that they are very different. Anatomically, the large numbers of trigeminal fibers, relative to the other sense organs, are impressive. Prutkin et al.[46] report that 75% of the input of the fungiform papillae comes from the trigeminal neurons that surround them. Even the taste bud itself seems organized to provide trigeminal information. Essentially, three major physiological classes of nerve fibers can be distinguished in the lingual branch of the trigeminal nerve:

* *Mechanorectors*, that mediate textural information
* *Thermoreceptors*, for discriminating temperatures
* *Nociceptors*, involved in the registration of pain and irritation

The role of the mechano- and thermoreceptors is clear. Thermoreceptors mediate thermal information and mechanoreceptors are designed to register texture. The appreciation of food is strongly related to its texture. Texture is defined as the sensory and functional manifestation of the structural, mechanical, and surface properties of foods (and drinks) detected through the senses of vision, hearing, touch, and kinesthetics.[47] They are also referred to as the somatosensory properties (see Figure 2.8).

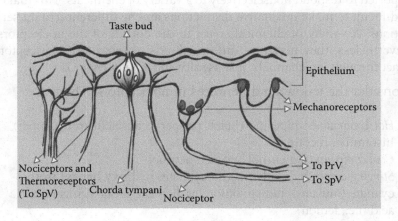

FIGURE 2.8   THE TRIGEMINAL CONNECTIONS IN THE MOUTH: PRINCIPAL TRIGEMINAL NUCLEUS (PRV) AND SPINAL TRIGEMINAL NUCLEUS (SPV). (FROM KATZ, D. B., NICOLELIS, M. A. L., SIMON, S. A., *AMERICAN JOURNAL OF PHYSIOLOGY: GASTROINTESTINAL AND LIVER PHYSIOLOGY*, 278, G6–G9, 2000. WITH PERMISSION.)

In the section "Personal Differences" in this chapter, it was shown that the number of fungiform papillae varies from person to person, and that this leads to different perceptions. If these papillae are also linked to the trigeminal system, it is not surprising that super-tasters experience oral irritants and other mouthfeel-related aspects more intensely than those with fewer fungiform papillae do. Viscous substances such as dairy fats, salad dressing, guar gum, and canola oil produce the most intense sensations to super-tasters. Medium-tasters and super-tasters could discriminate differences in fat content (10% or 40% fat) of salad dressings that the non-tasters could not. Although medium- and super-tasters showed no preference for either dressing, the non-tasters preferred the 40% fat sample. This implies that fat perception and preference can be linked to genetic and anatomical variation between individuals.

THE VERSATILITY OF NOCICEPTORS

Pepper is generally considered to be a taste and menthol a smell, even though they are actually examples of oral and nasal chemical irritations, transmitted principally by nociceptors. The focus in research is mostly on the variety and function of the nociceptors. Trigeminal neurons are reported to respond nonselectively to a variety of chemicals. This makes it difficult to make qualitative distinctions among different irritant sensations. A variety of chemicals react to the outside of the nociceptors; nevertheless, they may give different kinds of sensation. Nociceptors react through ion channels, just as salty and acidic flavors do.

Consider the following differences in nociceptors. There are:

- *Hot compounds*: piperine (black pepper), capsaicin (red pepper), histamine, nicotine
- *Cool compounds*: menthol and neomenthol
- *Stinging compounds*: allyl isothiocyanate, hydroxy benzyl isothiocyanate (mustard, radish, onion, garlic), zingerone (ginger), citric acid (i.e., lemon)
- *Burning compounds*: ethanol, cinnamic aldehyde (cinnamon), curcumin (turmeric, an essential ingredient of a curry)
- *Astringent/drying compounds*: hydrolyzable tannin and condensed tannin (wines)

The human reaction to oral and nasal irritants, such as piperine, capsaicin, menthol, and CO_2, has several properties that are different from gustation and olfaction. In the first place, oral and nasal irritants take longer to develop and tend to linger longer. Secondly, some irritants are known to desensitize during prolonged exposure. This is important in understanding how some cultures respond differently to spiciness. Thirdly, the responsiveness to irritation from capsaicin is not limited to the tongue. The throat and the roof of the mouth also contain nociceptors.[48,49]

ASTRINGENCY AND TRIGEMINAL SENSITIVITY

The human flavor registration is blurred through the different systems. This blurring is perhaps worst in the trigeminal sensations of astringency. The flavor of wines, especially young red ones, can be characterized by proanthocyanidins (tannins). Principally, tannins in foods must be considered as chemical stimuli (tannic acid), and yet the astringent sensations they produce seem largely tactile. Their specific flavor characteristic is described as astringent. Astringency is a trigeminal sensation that has a drying influence in the mouth. The term *astringent* is derived from the Latin *adstringere* or "binding." This effect is ascribed to the binding of salivary proteins and mucopolysaccharides (the slippery components of saliva), causing them to aggregate or precipitate, thus taking away the ability of saliva to coat and lubricate oral tissues. The mouth feels rough and dry, lips and cheeks seem to contract, and bluish-red strands of coagulated saliva are present on the tongue. Although chemical and physiological analysis would categorize astringency as a group of chemically induced oral tactile sensations, most wine tasters would say that astringency is an important component of wine flavor. Acidic polysaccharides in wine reduce the coarse, dry, and puckery perception of astringency. Perception is also influenced by ethanol content. This highlights, once again, the integrative nature of flavor in combining inputs from multiple modalities.[50,51]

Given all this, mouthfeel is a useful general term to cover all the effects obtained by the trigeminal system and registered orally and/or nasally. There is growing evidence that gustatory and trigeminal pathways interact. Gustatory fibers also respond to thermal and mechanical stimuli. The traditional basic flavors such as salty, sweet, and sour are not purely gustatory; they are felt as well. Just as in the case of the olfactory and gustatory systems, the trigeminal system is closely related and easily integrated into the other systems involved in flavor registration, which again makes it hard for any individual to differentiate among systems.[52,53]

THERE IS MORE THAN
MEETS THE MOUTH

SENSORY THRESHOLDS AND TIME INTENSITY

The presence of substances in foods and drinks does not necessarily mean that they are perceived. They need to be above what is called the sensory threshold. This is formally defined as the smallest amount of energy needed for a person to consciously detect a stimulus. Furthermore, there is the issue of "time intensity." It is slightly different but also influential. This refers to the moment that a substance is starting to be perceived. Some substances are perceived very quickly while others need time to develop. There may also be differences in the time that they are tasted. Some are gone quickly, while others linger for some time. Take the perception of acidity, for example. Some acids are perceived almost immediately, while others need some time to develop. This same phenomenon is also known in the perception of peppers. Cayenne peppers, especially, take some time to build up. This is illustrated in Figure 2.9. At Time 1, mainly substance A is tasted, together with B. Substance A is rather quickly gone, while the intensity of B gets stronger. At Time 2, the intensity of B is at its peak, and A and C are at par. At Time 3, the flavor is dominated by C and B.

These elements of tasting are important to realize. Flavor is a complex mix of all kinds of elements. Even if not all of them are detected, they may be there. It is one of the challenges of food pairing: keeping the undesired elements subthreshold or acceptably low and enhancing the desired elements. In a well-made dish of a very good chef, all elements fit together, including the ones that are not immediately tasted.

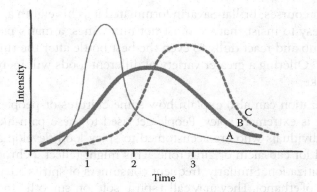

FIGURE 2.9   SENSORY THRESHOLD AND TIME INTENSITY.

Similarly, elements that may turn a wine faulty are mostly naturally present but at insufficient concentrations to adversely affect it. In fact, some of these concentrations may in effect have a positive effect on the wine flavor. An interesting example is *Brettanomyces*, a species of naturally occurring yeast often colloquially referred to as "Brett."[54] It lives on the skins of fruit. When *Brettanomyces* grows in wine, it produces several flavor compounds that arguably have a positive effect on wine, contributing to complexity and giving an aged character to young red wines. However, when the concentration of these compounds greatly exceeds the sensory threshold, the flavors and aromas that the wine should be expressing are dominated. The wine is less appealing and sometimes undrinkable; its quality is reduced.

ADAPTATION AND (DE)SENSITIZATION

The human senses have a special capacity: they can adapt to a certain stimulus. Sensitivity is influenced as a result of a prolonged exposure to the stimulus. Think of how one gradually sees more in the dark or how we adapt to bad smells until we do not notice them anymore. This last phenomenon is referred to as desensitization. It is a functional capacity enabling the organism to respond to changes in stimulation.[55] In eating and drinking, this happens all the time. It is one of the motivations to have several courses in a menu or to change a wine after courses. Brillat-Savarin formulated it as his eighth aphorism: "it is heresy to insist that we must not mix wines: a man's palate can grow numb and react dully to even the best bottle after the third glass from it."[56] Offering a greater variety of different foods will increase the total intake.

Desensitization can also explain how some cultures or people can eat food that is extremely spicy. People get used to these pain-like irritations. Individuals who are accustomed to spicy food develop a higher threshold for capsaicin or zingerone. This might reflect a chronic state of desensitization. Similarly, frequent consumers of spirits adapt to the oral burn of ethanol. They may call a spirit "soft" or "smooth." Infrequent consumers of the same drink are likely to experience it quite differently. The preference for these kinds of effects is likely to be acquired in most cases and will lead to individual assessments of sensory characteristics. Salt and acids are known to increase the effect of irritants.[57,58]

Adaptation is also well known in chemical flavors such as salt and sweetness. People grow accustomed to certain levels. In case of a sudden change in dietary habits (sugar or salt free), foods are likely to be judged as less palatable. After the flavor system is reorganized, appreciation increases. This adaptation generally takes about a week. It is possible that the regeneration of taste cells plays a role in this process.

Bitter flavors are considered to linger in the mouth, implying that perceived bitterness increases with successive intakes. This sensitization is compound specific. In one study, caffeine was reported to be the least affected by other bitter-tasting compounds tested, while—at the same time—it exerted the largest sensitization effect on the other compounds. This result is explained by the different ways bitter elements may be transmitted. Caffeine penetrates cell membranes to activate calcium channels directly. Such compounds may take longer to leave the gustatory cell than compounds that are registered otherwise.[59]

FLAVOR ENHANCEMENT OR SUPPRESSION

There is extensive research on the interaction of all kinds of sweet-, sour-, or bitter-tasting substances. When two different flavor components are mixed at moderate or strong concentrations, the mixture will often yield a flavor sensation that is less intense than the simple sum of the component flavors. This is called suppression.

Salt and umami are known as flavor enhancers and used to increase the flavor richness of all kinds of dishes. If salt is reduced in a product, the perception of other flavor components changes with it—even ones that are unrelated to salt. Likewise, sweetness or sourness will—for instance—enhance the fruity flavor of beverages or fruit.

Sweetness and sodium salts are known to suppress bitter flavors. Sugar in coffee decreases the bitter flavor of caffeine. On the other hand, the removal of sugar and fat from many foods unmasks underlying bitter attributes. Because sugar and fat both involve mouthfeel, it would be interesting to study the effects of the coating of papillae on the bitter perception. How salts and acids affect each other depends on concentration: they enhance each other at moderate concentrations but suppress each other at higher concentrations.

Chlorhexidine, a bitter-tasting antiseptic, severely impairs the identification of salty (NaCl, KCl) and bitter flavors (quinine HCl). Gymnemic acid (a mixture of bitter acidic glycosides) blocks the sweet flavor of all sweet stimuli. The same effect is observed with lactisole.[60,61,62]

Adding volatile aromatic compounds can compensate for the loss of some gustatory flavors. When a mixture of food odors such as strawberry, pineapple, raspberry, or caramel is added to sucrose it is judged sweeter than sucrose alone. The aromas of caramel and lychee are reported to suppress the sourness of citric acid.[63,64] Gustatory and odor stimuli interact most strongly when the associations are congruent and logical.

CO_2 stimulates not only trigeminal fibers but also gustatory ones, altering especially sweet, salty, and sour flavors. Adding CO_2 to a sucrose mixture reduced the relative proportion of sweetness, while sourness increased. Likewise, when added to a salty mixture, saltiness decreased, while the perception of sourness again increased. However, when carbonation was added to citric acid, its sourness decreased, while bitterness increased. CO_2 had little effect on the bitterness of quinine. CO_2 is also reported to block smell.[65]

CARRY-OVER EFFECTS

The effects mentioned up to now are mostly directly related senses. There are also carry-over effects: interference between bites and sips. The official definition states: the response to a stimulus is influenced by a previous sample. In everyday life, carry-over effects are inevitable and will also have a large effect. During the course of a meal or a tasting, previously tasted foods or drinks are likely to influence the perception of the next. There are several factors that are responsible for such interactions.[66]

The first phenomenon is called "successive contrast" and refers to the influence of flavor intensity. After having tasted something with a high flavor intensity, the intensity of the item following is perceived to be less intense than when it is preceded by something with a low flavor intensity.[67] A variant of successive contrast is "hedonic contrast," which can be best described by the Italian proverb *il meglio e l'inimico del bene*, which is translated as "the better is the enemy of the good." This implies that similar stimuli are rated less in quality when compared to something higher in quality. If tasted independently or compared to a lower quality stimuli, it is likely to be rated better.[68]

Another phenomenon is called "cross-adaptation." It is defined as a reduction of sensation of one stimulus following a prolonged stimulus by another. It is hypothesized that it is caused by shared receptor processes.[69] The use of alcohol together with food is another example of how the flavor of one substance may change the flavor of another. Moderate use of ethanol is reported to alter the registration of sensory properties of items consumed at the same time. Most reports mention a marked reduction of bitter flavors and slightly diminished ratings for sourness and saltiness, although sweet flavors are often slightly enhanced in combination with ethanol.[70]

Some proteins (such as in a soft-boiled egg), fat (butter, cream), and dissolved sugars (such as in honey) have the capacity to coat the mouth with a thin layer that will also influence the registration of succeeding sips or bites. We propose to call these kinds of influences residual flavors. The effect can easily be demonstrated by alternately tasting honey and fresh orange juice. After the honey, the orange juice will appear to be much more acidic. Likewise, acidity is likely to be enhanced by viscous residues of cream sauces. Literature on this subject is scarce. The capacity of substances to dissolve in water plays a role. Some substances do not mix and are called hydrophobic, others can and are called hydrophilic. Lipids in principle do not mix with water, and this influences perception of flavors that are dissolved in water.

Lipids are capable of reducing the oral burn of capsaicin and the bitterness of caffeine. Examples are the use of cream in coffee and of milk in tea. Chocolate is a beautiful example in itself. The difference between milk chocolate and plain chocolate is also explained by the use of fat and sugar (originally, just milk). As the cacao content rises, the flavor becomes more bitter and dry and the texture gets harder.

The results of a study on the influence of fat residues on the registration of other flavors are shown in Table 2.1. The intensity of sweetness, bitterness, and astringency decreased considerably, while the perceived intensity of salty and acidic increased. Differences in the chemical structure of these flavors may explain such effects and especially their inability to mix with water.[71]

TABLE 2.1
Effect of Oil on Flavor Intensity (% of the Scale)

Flavor	Acceptability before Oil	Intensity before Oil	Acceptability after Oil	Intensity after Oil
Sweet	34	47	40	10
Salty	77	69	78	77
Acidic	28	47	45	55
Bitter	26	68	23	43
Astringent	35	77	22	44

The influences mentioned here shed a particular light on the relevance of laboratory experiments in relation to real life situations. Flavor adaptation is frequently observed in research settings. However, during normal eating the results are quite different. Real life situations are almost by definition uncontrolled and, therefore, accidents and serendipities are the rule rather than the exception.[72]

INTRINSIC AND EXTRINSIC COMPONENTS OF FLAVOR

It will have become clear that tasting is an intricate process in which the gustatory, olfactory, and trigeminal systems work closely together. The elements that are registered are the *intrinsic components* of flavor. This implies that there are extrinsic components as well; external elements that are not taken in but nevertheless influence tasting. In this section, the role of these extrinsic factors is discussed—essentially how seeing and hearing are also involved in tasting. The role of these senses may be more important than often thought. They are not a part of the physical product but yet are strongly related to it. They serve as an introduction and help in selecting the right food products. At the same time, seeing and hearing can be potent drivers of liking and disliking. As such, these senses have a supportive function with regard to taste, which involves flavor appreciation. Packaging, brand names, health claims, and product images are examples of extrinsic factors that are known to influence flavor registration. The role and importance of these extrinsic factors should not be underestimated. More and more, researchers are interested in the effects of these factors and how they interact with the perception of intrinsic product characteristics.[73]

To start with, it may safely be hypothesized that our eyes and nose innately have an important warning function. We expect that food and drinks will keep us alive, to say the least. We need the nutrients to stay healthy and as sources of energy; yet milligrams or even less of certain toxics can be deadly. Our external senses guide us in the first selection. There are many examples that show that flavor registration is influenced by what we see. Of the various appearance characteristics of foods, color is probably the most important, especially if quality is related to the color. This is, for example, the case in the ripening of fruit or the association of color change with deterioration and spoilage; the color gives an indication of ripeness. Think also about how shades of brown give an indication of the level of frying and baking. Colors are also reported to influence the perception of refreshing and of sweetness. Red-colored yogurt will be judged as being fruitier or sweeter than uncolored yogurt, even though the level of sweetness is the same.[74] Conversely, green is associated with sour. In general, color influences taste thresholds, sweetness perception, pleasantness, and acceptability. Good-looking food will induce eating, just as seeing people enjoy a certain food will do so. Image or reputation can also have an effect, either positively or negatively.[75]

Research involving colors gives a fascinating insight into the functioning of our brain. In a French study, a white wine was artificially colored red, and subsequently wine experts olfactorily described it as a red wine. Apparently, in seeing things, we open virtual drawers in our brain and start looking for answers.[76] In other words, expectancy is a forceful driver of sensory perception. Not only colors lead to certain expectancies; packaging and labeling can also be a source of influence. A round, circular form is associated with round and easy flavors, while square forms are more readily associated with robust flavors. The bottle form of Burgundy in contrast to the one of Bordeaux is therefore well chosen and in correspondence with the general flavor of the wines of these regions. In a totally different research project, we experienced more of the same.

Visual or auditory health claims, such as fat-free, low-fat, diet, or sugar-free, do have influence. When tested blindly, subjects may like such a product less than the regular version, but when they know the difference, they may prefer it anyway. Price is another factor that can influence purchase behavior. Low prices are likely to be associated with inferior quality, while a high price may induce consumers to prefer a less liked product. Visual advertisements in general are designed to enhance taste qualities. Strong brand names are known to have a positive impact on flavor registration. These kind of strong relations have important implications for food design and packaging.[77]

Hearing contributes directly to flavor registration as well. In chewing, foods break and crack, and this sound is picked up by our ears. Think of crispness, crunchiness, and crackliness. The difference in perception between crunchy and crackly products and crispy ones can be explained by sound: crunchy and crackling sounds are lower than crispy ones. If a crisp product does not produce the expected sound, it is considered to be stale and of poor quality. These words should therefore be used with care as expectancy is involved. It would be a mistake to call a crunchy product "crispy."[78]

To conclude: in understanding the physiology of tasting, it is important to consider the role of extrinsic factors. They may have a strong influence on flavor registration and deserve a better qualification than "nonsensory variables" (as they are sometimes referred to in scientific literature).

SYNESTHESIA AND OTHER FORMS OF SENSORY INTERACTION

Extrinsic factors may influence the perception of the intrinsic qualities. In other words, the separate nerve systems (through which flavor is perceived) interact. One sensory system can evoke effects in another system. The general word for this phenomenon is *synesthesia*, which is defined as an involuntary conscious sensation that is induced by a stimulus in another modality. As the flavor system relies on an active involvement of different senses, synaesthesia-like, cross-modal effects are abundant. Sensory interactions occur every time we eat or drink. The flavor of individual components can be enhanced or suppressed. In order to understand the full extent of gastronomy, it is imperative to be aware of such phenomena.

Gustatory and olfactory interactions occur each time we eat or drink, and the importance of smell to flavor can hardly be underestimated. In general, the aroma of food can have a great influence on appreciation and palatability. The pleasing smell of freshly baked bread, just as that of baking cookies and cakes, roasting meat, coffee beans, or almonds, is the result of the Maillard reaction—a complex chemical reaction that influences the color, flavor, texture, and nutritional value of foods. Its flavor component is sought after, for it is known to be generally liked and appetizing.[79]

There is also an interesting cross-modal interaction between the gustatory and the trigeminal systems. Temperature has an especially strong influence on flavor perception. Melted ice cream tastes much sweeter than it does when cold. Warm beer tastes more bitter. In contrast, salty flavors seem to be enhanced at lower temperatures. Cooling and irritating compounds are reported to have various effects on gustatory flavors as well.[80]

It is interesting to see how smell "needs" other senses. We may have a hard time identifying a smell blindly; it gets much better when seeing the right color or with other sensory information. One study showed that smells that match the right color are perceived as being more intense than when in combination with another color. Good examples are lemon-yellow, strawberry-red, and caramel-brown. Specific brain areas showed higher activity if odors were paired with the matching colors, implying (for example) that caramel has a stronger odor when seen with the color brown rather than with other colors. Similarly, the auditory cortex is more active if a subject hears and sees someone talking. Such strong relations have important implications for food design and packaging.[81,82]

ARE PROFESSIONALS ABLE TO TASTE BETTER THAN AMATEURS?

This chapter demonstrates how multisensorial tasting really is, including its intrinsic and extrinsic components. One question to consider is whether professional tasters in whatever field are actually better equipped to register flavor than amateurs. The answer is no. Of course, whether amateur or professional, the basic systems need to work properly so that flavor is actually registered as it should be. The difference between the two relates language and experience. If a professional taster takes a sip or bite, the language area in the brain is active from the beginning. The professional is likely to be asked his opinion. The amateur just tastes and starts looking for words later, if asked. The latter taster may register the same but is way behind in naming what is being tasted. And then there is also experience. Professional tasters are trained in their field. They have learned to label the elements, often using specific words and jargon that have little or no meaning to nonprofessionals.

Everyone tastes within a personal framework made up by biological, psychological, social, cultural, and economic forces. Chapter 7, on food appreciation and liking, will elaborate on these factors. For now, we may safely conclude that there is no way of knowing exactly how someone perceives a certain dish or drink. Sensory capacity will differ from person to person; furthermore, flavor registration is influenced by situational circumstances, including culture, beliefs, experience, preferences, and so forth. There is a lot more to be discovered. Nevertheless, you can learn to recognize a wine that is true to its character, just as with anything else that we eat or drink. Perception may be personal but only to a certain extent and more in the field of appreciation rather than when it is about identification.

At the start of this chapter we raised the issue of whether you would expect a chapter on the working of our eyes in a book about color. This chapter shows how functional it is to consider flavor as a product characteristic. The psychological construct that we make from it is easily influenced by many factors. In the next chapter, we will focus on products and their physical flavor (see Figure 2.10).

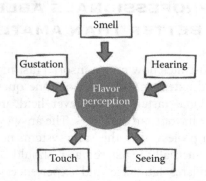

FIGURE 2.10   FLAVOR PERCEPTION—THE RESULT OF AN INTRICATE COOPERATION OF THE HUMAN SENSES.

SUMMARY

Flavor can be studied from a human and a product angle. In this chapter, the human angle has been reviewed: tasting or flavor registration. Tasting does not rely on a single sense but is a sublimation of the human senses. Three different sensorial systems are directly involved in the registration of the so-called intrinsic flavor components: the gustatory, olfactory, and trigeminal systems. Each of these systems seems to work independently to convey information to brain areas where it is integrated and assembled to perceive flavor. Differences between humans and cross-modal sensorial interferences influence tasting just as the extrinsic factors that we see, hear, or smell externally do before we take a sip or a bite. If the physiology underlying the transduction of single sensory modalities is already complex, the physiology underlying a multisensorial experience such as tasting is at least three times as complex.

The four basic tastes hypothesis has dominated the field and research methods. Attention has been too focused on the gustatory part of flavor registration. Consequently, answers on flavor registration as a whole are not to be expected (and were not found); gustation is "just" one part of flavor perception and is influenced by trigeminal and olfactory elements. Furthermore, the brain may have a top-down influence on perception. Also, other sensory systems such as sight and sound may influence flavor. This implies that real foods studied in naturalistic settings are likely to provide better answers on flavor registration in general.

There is no doubt that tasting is a subjective affair. Everyone tastes within what may be considered a personal frame made up by biological, psychological, social, cultural, and economic forces. There is no way of knowing exactly how somebody perceives a certain dish or drink. Sensory capacity differs from person to person. In addition, flavor registration is influenced by situational circumstances, including culture, beliefs, experience, preferences, and so forth.

There is a huge pitfall that has trapped most people who discuss flavor or want to grasp its versatility: people are led to believe that it is no use to even try describing the structure of flavor, or classifying it and using it—as if personal visual or auditory problems would prevent one from studying light or sound. Flavor can be studied; it does have an objective side, and it does help to view it as such. In wanting to know more about color, it is quite logical to study the physical aspects of color and not the human perception of light. Likewise, potential answers on flavor are likely to be found at the physical product level and hardly at the sensorial level of flavor registration. So do not get trapped by the subjective side; enjoy the possibilities that the other side of flavor holds in store. The next chapter will elucidate the physical side of flavor.

NOTES

1. ISO (International Organization for Standardization) (1992). *Standard 5492: Terms related to sensory analysis.* Geneva, Switzerland: ISO.
2. Delwiche, J. (2003). *Attributes believed to impact flavor. An opinion survey.* Journal of Sensory Studies 18:347–52.
3. Shepherd, G. (2012). *Neurogastronomy. How the brain creates flavor and why it matters.* New York: Columbia University Press.
4. Western Center for Food Safety (WCFS) Food Summit (2002). *Food texture: perception and measurement. Report of an international workshop.* Food Quality and Preference 13:237–55.
5. Wilkinson, C., Dijksterhuis, G. B., Minekus, M. (2001). *From food structure to texture.* Trends in Food Science & Technology 11(12):442–50.

6. Guinard, J-X., Mazzucchelli, R. (1996). *The sensory perception of texture and mouthfeel.* Trends in Food Science & Technology 7:213–19.
7. Guinard, J-X., Mazzucchelli, R. (1996). *The sensory perception of texture and mouthfeel.* Trends in Food Science & Technology 7:213–19.
8. Gilbertson, T. A., Damak, S., Margolskee, R. F. (2000). *The molecular physiology of taste transduction.* Current Opinion in Neurobiology 10:519–27.
9. Beidler, L. M. (1978). *Biophysics and chemistry of taste.* In Carterette, E.C. and Friedman, M. P. *Handbook of perception,* Volume VIA, Tasting and Smelling (pp. 21–49). New York: Academic Press.
10. Chou, H.-C., Chien, C.-L., Huang, H.-L, Lu, K.-S. (2001). *Effects of zinc deficiency on the vallate papillae and taste buds in rats.* J. of the Formosan Medical Association 100:326–35.
11. Miller, I. J. Jr, Reedy, F. E. Jr. (1990). *Variations in human taste bud density and taste intensity perception.* Physiology & Behavior 47:1213–9.
12. Temple, E. C., Hutchinson, I., Laing, D. G., Jinks, A. L. (2002). *Taste development: differential growth rates of tongue regions in humans.* Developmental Brain Research 135:65–70.
13. Prutkin, J., et al. (2000). *Genetic variation and inferences about perceived taste intensity in mice and men.* Physiology & Behavior 69:161–73.
14. Schiffman, S. S., Graham, B. G., Sattely-Miller, E.A., Peterson-Dancy, M. (2000). *Elevated and sustained desire for sweet taste in African-Americans: a potential factor in the development of obesity.* Nutrition 16:886–93.
15. Tepper, B. J., Christensen, C. M., Cao, J. (2001). *Development of brief methods to classify individuals by PROP taster status.* Physiology & Behavior 73:571–7.
16. Birch, L. L. (1999). *Development of food preferences.* Ann. Rev. Nutrition 19:41–62.
17. Yackinous, C. A., Guinard, J.-X. (2002). *Relationship between PROP (6-n-propylthioracil) taster status, taste anatomy and dietary intake measures for young men and women.* Appetite 38:201–9.
18. Bartoshuk, L. M. (2000). *Comparing sensory experiences across individuals: recent psychophysical advances illuminate genetic variation in taste perception.* Chemical Senses 25:447–60.
19. Tepper, B. J., Nurse, R. J. (1998). *PROP taster status is related to fat perception and preference.* Annals New York Academy of Science 30(855): 802–4.
20. Hänig, David P. (1901). *Zur psychophysik des geschmackssinnes.* Philosophische Studien 17:576–623.
21. Smith, D. V., Margolskee, R. F. (2001). *Making sense of taste.* Scientific American 288:26–33.

22. Lindemann, B. (1999). *Six common errors: criticism of the human tongue maps as seen in some textbooks*. www.med-rz.uni-sb.de/med-fak/physiol2/LDM/chemotopic_1.htm.
23. Bellisle, F. (1999). *Glutamate and the UMAMI taste: sensory, metabolic, nutritional and behavioral considerations. A review of the literature published in the last 10 years*. Neuroscience and Biobehavioral Reviews 23:423–38.
24. Bray, G. A. (2000). *Afferent signals regulating food intake*. Proceedings of the Nutrition Society 59:373–84.
25. Rolls, E. T., Critchley, H. D., Browning, A. S., Hernadi, I., Lenard, L. (1999). *Responses to the sensory properties of fat of neurons in the primate orbifrontal cortex*. Journal of Neuroscience 19:1532–40.
26. Cubero-Castillo, E., Noble, A. C. (2001). *Effect of compound sequence on bitterness enhancement*. Chemical Senses 26(4):419–24.
27. Smith, D. V., St. John, S. (1999). *Neural coding of gustatory information*. Current Opinion in Neurobiology 9:427–35.
28. Cowart, B. J. (1998). *The addition of CO_2 to traditional taste solutions alters taste quality*. Chemical Senses 23:397–402.
29. Schiffman, S. S., et al. (2000). *Effect of temperature, pH, and ions on sweet taste*. Physiology and Behavior 68:469–81.
30. Trivedi, B. (2012). *Gustatory system: the finer points of taste*. Nature 486:S2–3.
31. Doty, R. L. (2001). *Olfaction*. Ann. Rev. Psychology 52:423–52.
32. Delwiche, J. (2003). *Attributes believed to impact flavor. An opinion survey*. Journal of Sensory Studies 18:347–52.
33. Brand, G., Millot, J.-L., Henquell, D. (2001). *Complexity of olfactory lateralization processes revealed by functional imaging: a review*. Neuroscience and Biobehavioral Reviews 25:159–66.
34. Freeman, W. J. (1991). *The physiology of perception*. Scientific American 264:78–85.
35. Bell, G. A. (1996). *Molecular mechanisms of olfactory perception: their potential for future technologies*. Trends in Food Science & Technology 7:425–31.
36. Rozin, P. (1982). *Taste-smell confusions, and the duality of the olfactory sense*. Perception & Psychophysics 31:397–401.
37. Monell Chemical Senses Center (2001). *Monell connection*. Spring 2001.
38. Birch, L. L. (1999). *Development of food preferences*. Ann. Rev. Nutrition 19:41–62.
39. Doty, R. L., (2001). *Olfaction*. Ann. Rev. Psychology 52:423–52.
40. Rolls, B. J. (1999). *Do chemosensory changes influence food intake in the elderly?* Physiology & Behavior 66:193–7.

41. Finkelstein, J. A., Schiffman, S. S. (1999). *Workshop on taste and smell in the elderly: an overview.* Physiology & Behavior 66:173–6.

42. Murphy, C., Nordin, S, Jinich, S. (1999). *Very early decline in recognition memory for odors in Alzheimer's disease.* Aging, Neuropsychology, and Cognition 6:229–40.

43. Rolls, E. T., Critchley, H. D., Browning, A. S., Hernadi, I., Lenard, L. (1999). *Responses to the sensory properties of fat of neurons in the primate orbifrontal cortex.* Journal of Neuroscience 19:1532–40.

44. Guinard, J-X., Mazzucchelli, R. (1996). *The sensory perception of texture and mouthfeel.* Trends in Food Science & Technology 7:213–9.

45. International Organization for Standardization. (1992). *Standard 5492: Terms related to sensory analysis.* Geneva, Switzerland: Author.

46. Prutkin, J., Duffy, V. B., Etter, L., Fast, K., Gardner, E., Lucchina, L. A., Snyder, D. J., Tie, K., Weiffenbach, J., Bartoshuk, L. M. (2000). *Genetic variation and inferences about perceived taste intensity in mice and men.* Physiology & Behavior 69:161–73.

47. Szczesniak, A. S. (2002). *Texture is a sensory property.* Food Quality and Preference 13:215–25.

48. Green, B. G. (1996). *Chemesthesis: pungency as a component of flavor.* Trends in Food Science & Technology 7:415–9.

49. Dessirier, J.-M., Simons, C. T., O'Mahony, M., Carstens, E. (2001). *The oral sensation of carbonated water: cross-desensitization by capsaicin and potentiation by amiloride.* Chemical Senses 26:639–43.

50. Gawel, A., Iland, P. G., Francis, I. L. (2001). *Characterizing the astringency of red wine: a case study.* Food Quality and Preference 12:83–94.

51. Lawless, H. T. (1996). *Flavor.* In M. P. Friedman & E. C. Carterette (Eds.), *Cognitive ecology* (pp. 325–380). New York: Academic Press.

52. Boucher, Y., et al. (2003). *Trigeminal modulation of gustatory neurons in the nucleus of the solitary tract.* Brain Research 973:265–74.

53. Carstens, E., et al. (2002). *It hurts so good: oral irritation by spices and carbonated drinks and the underlying neural mechanisms.* Food Quality and Preference 13:431–43.

54. The term *Brettanomyces* is Greek for "British fungus." The term originated when this yeast was found to be a cause of spoilage in English ales.

55. Dalton, P. (2000). *Psychophysical and behavioral characteristics of olfactory adaptation.* Chemical Senses 25:487–92.

56. Brillat-Savarin, J. A. (1826). *Physiologie du goût.* Paris: Sautelet. American edition: Fisher, M. F. K., trans., *The physiology of taste* (1949). New York: Knopf.

57. Prescott, J. (1999). *The generalizability of capsaicin sensitization and desensitization.* Physiology & Behavior 66:741–9.

58. Prescott, J. (2000). *Responses to repeated oral irritation by capsaicin, cinnamaldehyde and ethanol in PROP tasters and non-tasters.* Chemical Senses 25:239–46.

59. Walters, D. E. (1996). *How are bitter and sweet tastes related?* Trends in Food Science & Technology 7:399–403.

60. Walters, D. E. (1996). *How are bitter and sweet tastes related?* Trends in Food Science & Technology 7:399–403.

61. Breslin, P. A. S. (1996). *Interactions between salty, sour and bitter compounds.* Trends in Food Science & Technology 7:390–9.

62. Gent, J. F., Frank, M. E., Hettinger, T. P. (2002). *Taste confusions following chlorhexidine treatment.* Chemical Senses 27:73–80.

63. Prescott, J. (1999). *Flavour as a psychological construct: implications for perceiving and measuring the sensory qualities of foods.* Food Quality and Preference 10:349–56.

64. Stevenson, R. J., Prescott, J., Boakes, R. A. (1999). *Confusing taste and smells: how odours can influence the perception of sweet and sour tastes.* Chemical Senses 24:627–35.

65. Cowart, B. J. (1998). *The addition of CO_2 to traditional taste solutions alters taste quality.* Chemical Senses 23:397–402.

66. Cubero-Castillo, E. & Noble, A.C. (2001). *Effect of compound sequence on bitterness enhancement.* Chemical Senses 26:419–24.

67. Schifferstein, H. N., Oudejans, I. M. (1996). *Determinants of cumulative successive contrast in saltiness intensity judgments.* Perception and Psychophysics 58(5):713–24.

68. Zellner, D. A., Kern, B. B., Parker, S. (2002). *Protection of the good: subcategorization reduces hedonic contrast.* Appetite 38:175–80.

69. Stevens, D. A. (1996). *Individual differences in taste perception.* Food Chemistry 56:303–11.

70. Mattes, R. D., DiMeglio, D. (2001). *Ethanol perception and ingestion.* Physiology & Behavior 72:217–29.

71. Valentova, H., Pokorný, J. (1998). *Effect of edible oils and oil emulsions on the perception of basic tastes.* Nahrung 42:406–8.

72. Theunissen, M. J., Polet, I. A., Kroeze, J. H., Schifferstein, H. N. (2000). *Taste adaptation during the eating of sweetened yogurt.* Appetite 34:21–7.

73. Lange, C., Rousseau, F., Issanchou, S. (1999). *Expectation, linking and purchase behaviour under economical constraint.* Food Quality and Preference 10:31–9.

74. Zellner, D. A., Durlach, P. (2002). *What is refreshing? An investigation of the color and other sensory attributes of refreshing foods and beverages.* Appetite 39:185–6.

75. Clydesdale, F. M. (1993). *Color as a factor in food choice.* Critical reviews in Food Science and Nutrition 33:83–101.
76. Morrot, G., Brochet, F., Dubourdieu, D. (2001). *The color of odors.* Brain and Language 79:309–20.
77. Lange, C., Rousseau, F., Issanchou, S. (1999). *Expectation, linking and purchase behaviour under economical constraint.* Food Quality and Preference 10:31–9.
78. Duizer, L. (2001). *A review of acoustic research for studying the sensory perception of crisp, crunchy and crackly textures.* Trends in Food Science & Technology 12:17–24.
79. Martins, S. I. F. S., Jongen, W. M. F., Van Boeckel, M. A. J. S. (2001). *A review of Maillard reaction in food and implications to kinetic modeling.* Trends in Food Science and Technology 11:364–73.
80. Talavera, K., Ninomiya, Y., Winkel, C., Voets, T., Nilius, B. (2007). *Influence of temperature on taste perception.* Cellular and Molecular Life Sciences 64:377–81.
81. Shimojo, S., Shams, L. (2001). *Sensory modalities are not separate modalities: plasticity and interactions.* Current Opinion in Neurobiology 11:505–9.
82. Calvert, G. A., Campbell, R., Brammer, M. J. (2000). *Evidence from functional magnetic resonance imaging of crossmodal binding in the human heteromodal cortex.* Current Biology 10:649–57.

CHAPTER 3

Understanding Flavor

FINDING WORDS: COMPARING
APPLES TO ORANGES

This chapter focuses on flavor as a concept and the theoretical flavor styles model. After we have explained what flavor is in general terms, we move into practice; the flavor of beverages and foods is elaborated upon in more detail in Chapters 4 and 5, respectively.

By now you have become familiar with tasting. All our senses work together to register flavor. Everything that our senses record is converted into electrochemical nerve impulses, which make up the common language of the brain. When the signals reach specific areas in the brain, something needs to be done. Registering a signal is one thing; giving it a name is quite another matter. And this is not a sensorial, but a brain task. The words that we use to address what we have experienced are human inventions. To illustrate: salt is a natural substance; it can be derived from the sea or from ancient salt banks deep under the ground. We have learned to recognize this mineral through our senses. Chemically, it is sodium chloride (NaCl), and the official geological name of the mineral is halite (named by E. F. Glocker in

1847). Salt content can be measured, and the effects of adding salt to something can be studied. The name salt is used all over the world. This is extremely functional because it enables effective communication. The specific words we use to describe physical phenomena are merely names chosen at a certain time. However, once a decision is made about what something is called, it is imperative that everybody uses that term consistently and, if possible, worldwide. The most useful concepts are well defined, unambiguous, easily recognized and understood, and value free.

However trivial and obvious this seems, it highlights the problem with describing flavor: the proper words are missing. This is best illustrated by asking people about the flavor of something they have just eaten or drunk. It is highly likely that they will tell you how yummy, delicious, disgusting, filthy, acceptable, or whatever it was. They are likely to share a personal judgment rather than giving a useful description of the flavor. This has become so common that we have all accepted the saying "there's no arguing about taste." And every one of us is excused for doing so: our vocabulary is poor; useful, descriptive words are lacking; or they are poorly defined or understood. And certainly these descriptive words have not been universally accepted. The ambition of this book is to turn that situation around.

Suppose we do what another expression suggests we had better not do: compare apples to oranges. This comparison is considered unfair, but as far as I am concerned, that depends on how you look at it. There is no reason whatsoever why these fruits cannot be compared. Clearly, they are very different, and everyone would agree on the differences. There is no reason to argue; the differences are facts, and they can be assessed objectively. The most apparent difference is texture. The apple is crisper and has a bite, while the orange is juicy. The aromas are quite different, but both tend to be sweet and acidic. In general, most apples will also be more acidic, while oranges are likely to be sweeter. Delving deeper, one runs into all kinds of differences. Not all apples and oranges are created equal. Within the apple family, a green apple, such as a Granny Smith, will be more crisp and acidic than a reddish apple such as a Cox Orange, which is sweeter and softer. Differences in ripeness could also easily give different outcomes. The riper the fruit, the sweeter the taste is the general rule. However, the market dictates earlier harvesting—before fruits have reached maximum maturity and when the flesh is still firm. They can be kept longer, and there is less bruising. It is therefore easier, and in some ways better, to address the differences between a specific apple and a specific orange. Some people prefer apples, and others like oranges better. Or, some like oranges better one day and apples another. To conclude, please note that as soon as you try to describe the flavor of an apple or an orange just using the so-called basic taste, you are in trouble. As can be expected, this general concept is not only faulty for explaining how tasting works, it also is inadequate for describing flavor.

WORDS DO NOT COME EASILY

So we need words, or rather we need to define concepts. Finding them has been a demanding, interesting, and even rewarding undertaking. It may seem that this started out as a well-defined mission. But this is not the case. You may even say that we stumbled over the words while conducting another type of research. We wanted to know more about the flavor of wine in particular, to get a better understanding of the combination of wines and foods.

In 1989, together with Dr. Ir. Cramwinckel of Wageningen University, a control group of twenty wine lovers was formed. Over the course of a year, they convened for a series of tastings at which different series of wines were tasted and the differences assessed. What makes one wine taste different from another? Advanced statistical techniques (principally free choice profiling) were used to interpret the tasting notes. Free choice profiling is divided into two components. In the first part, subjects use their own words (free choice). With factor analysis, these words can be clustered—although the words may be different, they must have something in common. The group is then asked to find words (labels) that adequately describe each cluster. The second part of the research beings once a consensus is reached. The labels are then evaluated to determine if they indeed can be used to adequately indicate differences in wines (profiling). In this process, labels are discarded or redefined if necessary. This study marked the birth of the universal flavor factors: mouthfeel, flavor intensity, and flavor type.

As we now know, these terms proved not only to be very useful to describe the flavor of wines, but foods and other drinks could also be classified and analyzed using these terms. All of a sudden interactions between wines and foods could be explained. These terms are simple, easily applicable, value free, and universal.

UNIVERSAL FLAVOR FACTORS

Mouthfeel, flavor intensity, and flavor type are the *universal flavor factors* that help to describe chocolate and chips, French fries and mayonnaise, soft drinks and milk, beer and wine, coffee and tea, meat and vegetables. In other words, in everything we could possibly taste, these same flavor factors can be distinguished but to differing degrees. The challenge is to be able to recognize these factors. An important advantage is that most people perceive them in the same way. These factors will now be explained one by one.

MOUTHFEEL: TYPES

Mouthfeel is the feeling a product gives in the mouth, whether it is a food or a beverage. We distinguish two basic dimensions of mouthfeel: *contracting* and *coating*. Each of these concepts has its own specifics.

CONTRACTING: THE BASIC FORM

Think of what you feel in your mouth when you eat a green apple, fresh lettuce, citrus fruits, tomatoes (the ones that are generally available in shops, not fully matured), leeks, turnips, all kinds of radishes, raw onions, chives, parsley, and—as an extreme—vinegar. Likewise, salt and all kinds of sensations that are in fact a kind of irritation (pepper, mustard, horseradish, ginger) cause contraction in the mouth—just as carbonic acid (CO_2) does in a variety of drinks, such as mineral water, sparkling wines, beer, and soda. Light, fresh white and red wines with a nice acidity are examples of "contracting" drinks. A low temperature also makes the mouth contract. This implies that serving temperature influences mouthfeel and flavor as a whole. In general, contraction gives the impression of refreshment, of cleansing the mouth. Contracting elements will often stimulate saliva flow.

COATING: FATTY/SWEET

The word *coating* is chosen for the characteristic of elements that leave a thin layer in the mouth. Think of how a spoonful of honey will make saliva thicker, more viscous. Fats, such as butter, cream, oils, and peanut butter, have the same effect, just like other sugars in solutions. You can imagine that syrup with a little water is much more viscous and coating than the same syrup with more water. Sweet wines such as Port and the so-called noble rot or botrytis wines are rich in sugar and therefore coating.

Besides fats and sugar, there is a third group that can be coating: proteins. Just imagine a soft-boiled egg and you have the best example. At the same time, you will realize how preparation influences mouthfeel. The same egg, cooked just for a few minutes longer, is no longer coating but dry, especially the yolk! The same thing happens when meat or fish has been cooked just a bit too long—instead of juicy and on the coating side, it has lost its juiciness. Some people like it that way, while others do not. That is why the waiter asks how you like your meat. Umami will be further elucidated later in this book. Nevertheless, it is important to mention here that umami refers to the flavor of a specific amino acid (glutamic acid), which is one of the important building blocks of proteins. It also has an effect on coating. Some examples are jelly, meat stocks, dried mushrooms, fermented sauces, cured hams, and wood fermented Chardonnay wines of which the lees have been stirred regularly. Generally people refer to umami as rich or savory.

Sugars, fats, and certain proteins may all be coating. Nevertheless, it is important to note that the effect of wines will not be the same. This has to do with chemical properties of the layer. Some of them, and especially fats, do not mix with water. They are hydrophobic and react adversely to wines. In other words: with regard to food combination with wines, it is important to distinguish the types of coating. Not all coats are waterproof, which is good to know—especially when you are having wine with your food. This will be discussed later in this chapter.

MOUTHFEEL: FROM NEUTRAL TO DRY

In mouthfeel, contracting and coating are quite opposite. Both can be scaled from low to high. All foods and beverages have contracting and coating elements to a certain extent. The starting situation can be called neutral and is neither contracting or coating. It may well be the condition Fernel had in mind in the sixteenth century when he proposed tasteless as a basic taste. Water, plain bread, or rice could be characterized as being neutral; they are neither coating nor contracting. Some of these elements have the capacity to absorb saliva. In many cases, this can be attributed to starch. Think of how rice, potatoes, or bread absorb all kinds of fluids, saliva included. This implies that neutral and dry are closely related. Dry is also neither contracting nor coating. A simple piece of bread neutralizes the mouth. You often see bread at wine tastings for that reason. Toast or crusts in general absorb as well but are not as neutral as the crumbs of bread. Just think of how the crust of freshly roasted meat, toast, or puff pastry enhances flavor, and also how its crispness influences our perception of quality.

Astringency is another form of dryness. It is mostly found in bitter substances that reduce the coating capacity of saliva. The tannin of a young red wine is a good example of this effect. Other examples include the bitterness of espresso coffee and some types of tea. Plain chocolate is also a beautiful example: as the cacao content gets higher, the chocolate gets dryer, in the bar as well as in the mouth. In the case of chocolate, dryness is directly related to high flavor richness.

FINDING BALANCE: MIXING NEUTRAL/ DRY, COATING, AND CONTRACTING

Real world foods and drinks are usually mixtures of contracting, dry, and/or coating elements. Ice cream is one common example. The basis is extremely sweet; it needs to be so to create very fine crystals that we find attractive. In the unfrozen form, it would be coating to the extreme. However, when the mixture is frozen and becomes ice cream, the mouthfeel changes. The cold makes it very contracting and masks the coating side.

Another type of mixture is found in a basic French dressing, vinaigrette. In this case, the oil (coating) and vinegar (contracting) find a balance, aided by mustard, salt, and pepper (all contracting). Types of oil differ in viscosity and therefore in coating, just as types of vinegar—and mustard—can be (very) different in acidity and thus in contracting. You need your tongue and personal preference to find the right mixture. It has become quite common to add sugar (coating) to the vinaigrette. It helps to mask the acidity and adds something on the coating side, but it is not the true recipe. If you find your vinaigrette too acidic, the logical solution would be to add more oil or to use a different type that is more viscous. It is also important to mention that balance is not necessarily found when the parts are equal. Most vinaigrettes are made up of least 70% oil.

And then there is pastry, especially puff pastry. Even though it is made with butter, the end result is definitely dry. The wonderful layered crust mixes beautifully with all kinds of creams that are rich and coating. This richness is affected by the dryness of the pastry; the balance gets better as it moves away from the fatty richness.

In wines and other drinks, balance is achieved by playing with the contracting elements acidity, carbonation, and temperature and with sweetness as a coating element. It is interesting to note that under the German wine law, white wines can be still be called *trocken* (dry) with nine grams of sugar per liter with a maximum of seven grams of acidity. This is generally called the "acidity plus two rule." Champagne can be called *brut* (dry) with a maximum of 13 grams per liter. In both of these cases, the sweetness is not noted; when tasted, both wines would be characterized as being contracting.

In these examples, balance is achieved by mixing substances. Balance can be observed within products as well. Think of how the crema cover of an espresso is coating, but the coffee itself is concentrated and bitter. Espresso lovers and baristas judge the quality of the brew by the thickness and color of the crema. Another illustrative example is found in olive oil, which is essentially coating. The best olive oils, however, have a spicy, peppery note that is contracting and makes quality oils not as viscous as the simpler and cheaper types.

FLAVOR INTENSITY

Flavor intensity is the amount of flavor. Flavor intensity is comparable to the volume of a sound system. If it is turned up, only the power or intensity changes, not the music. Likewise, there are products that plainly have more flavor than others. Let us take the apple and the orange again. However different they are, the flavor intensity of most oranges will be higher than that of most apples. It may seem strange to compare these fruits in this respect, but it can be done. Likewise, all kinds of products can be compared to each other. If you want to know if two wines are different in flavor intensity, just compare them next to each other. Take a sip of the first and a sip of the second, then immediately go back to the first one again. If the second wine is higher in intensity, the first wine will have less taste. But if the first wine has the higher intensity, it will be perceived as even richer than before. This is called a triangle test, and it is a useful technique to assess the flavor profile of many products.

It is important to note that a difference in flavor intensity does not say anything about quality. Just as the volume knob of the stereo does not say anything about the quality of the music. You may even take this a step further; there are many foods with a lower flavor intensity that are still much appreciated. Take poultry for example—its flavor intensity is lower than that of beef. Yet it is highly popular around the world. Products with a high flavor intensity can be rather dominating, pushing away interesting but fragile flavors. On the other hand, lower flavor intensity demands attention from the taster. Metaphorically speaking, flavors that are low in intensity do not come to you; you have to go to them. This means that a high flavor intensity is ideal in all those situations where you do not want to go to the trouble of tasting. This is exactly why the flavor intensity of successful fast foods is high.

Although we go into that in much more depth in Chapter 4, it is important to note that salt and umami components can be very effective in raising flavor intensity. Cooking techniques such as grilling, frying, and especially deep-frying are also known to raise flavor intensity. A piece of fish or chicken can easily illustrate that. Imagine it poached or steamed; its fragile flavor is well preserved, and the flavor intensity stays on the low side. Now fry it, and the flavor intensity goes up. Fry it even more or grill it and the intensity goes up further. It may even reach a point that the flavor of the preparation becomes more important than that of the primary ingredient: all you taste is the brown crust; it dominates the original flavor of the meat. These changes are the result of what frying does to certain molecules of the meat.

When combining wines and foods, the flavor intensity had better all be in the same league. A dish with a high flavor intensity will leave not much of the wine if it is one with an intensity that is markedly lower. In the same way, wines can rule over dishes and interfere with the efforts of the chef to make something special.

THE FLAVOR TYPE (FRESH OR RIPE FLAVOR TONES)

To conclude, flavor type can be assessed. Essentially, this is about aroma components. The complexity of naming aromas is greatly reduced by distinguishing just two types of flavor tones: *fresh* and *ripe*. However different, the flavor type of both apples and oranges can be classified as being fresh and distinctly different from pears, bananas, and melons. The flavor type of the latter three is rather ripe. Please note that, when used in this respect, the word *ripe* does not refer to the ripeness of the fruit: a beautiful ripe apple will always be fresh. You can also imagine berries, lemons, cucumber, fennel, mint, and parsley as being fresh. It is a flavor type that makes you think of spring and summer. Ripe flavor tones are those of the fall and winter. Examples are nuts, honey, caramel, mushrooms (especially dried ones), and spices such as vanilla, nutmeg, cinnamon, and the sultry aromas of musk and roses. Just to be clear: I'm not suggesting that there are only fresh or ripe flavor tones. The words *fresh* and *ripe* are meant to be used as indications of the type of flavor and to help give a better insight into the flavor profile.

Just as with flavor intensity, preparation will easily influence the flavor type. Looking at fish or poultry again—when fried and grilled, ripe flavor tones come out. With raw onions and bell peppers, the fresh flavor tones are abundant. However, when put in the oven and left to stew for at least half an hour, the fresh flavor tones are gone. The flavor tones have turned into ripe ones; they even become sweet. Potatoes are another example. Boiled or mashed they are still quite neutral, neither fresh nor ripe. The ripe flavor tones enter the scene when potatoes are fried.

In general, herbs and spices are very practical for giving direction to the flavor type. This is good to know if you want your dish to match a certain wine. If the wine has predominantly fresh flavor tones, add herbs that bring out the fresh flavor tones. If, on the other hand, the wine has aged in wood (for instance), causing it to be predominantly ripe, use riper herbs and spices as additives.

Finishing with a practical example, theoretically we will prepare a chicken breast with a creamy sauce and tarragon. How can we manipulate the flavor profile of the dish? Its mouthfeel is mainly influenced by the sauce. If the sauce is reduced more, it gets more coating. If we want it more on the contracting side we should, for instance, start with a reduction of some wine and shallots. When the cream is added, the level of reduction determines how coating the sauce will be. Flavor intensity is controlled by how much the chicken is fried. If we fry it a little too long, the mouthfeel of the chicken will turn dry. The tarragon in our sauce is added for the fresh flavor tones. If our wine is aged in wood, we might want to use rosemary or thyme instead or to add some mushrooms and/or garlic. This demonstrates that the flavor profile of any dish is not a given fact. It is in the hands of chefs, or winemakers and brewers in the case of wines and beers (see Table 3.1).

FLAVOR RICHNESS AND COMPLEXITY

In tasting notes or flavor descriptions, it is not unusual to see the word *complexity* used. Flavor richness and complexity are not necessarily the same. We have seen that flavor richness is a blend of flavor intensity and flavor type. Complexity has to do with how well the different flavors fit together. Complex flavors keep on surprising you. Every sip or bite will be slightly different. Less complex flavors may have a high flavor intensity, and also be well liked, but only for a short time; they will quickly bore the palate. They become "what you see is what you get." This is a very useful quality if you are busy or stressed and do not have the time to sit and relax and enjoy the complexity. Indeed, complex flavors demand a certain frame of mind or they are likely to be missed. In other words: there are a lot of circumstances where simple products are very appropriate and others where the complex ones fit better.

TABLE 3.1
Example of Fresh and Ripe

Examples (Not Limitative)	Fresh	Ripe
Fruit	apple, lemon, lime, red currant, strawberry, raspberry	banana, melon, pear, peach
Herbs/spices	tarragon, dill, chive, chervil, parsley, aniseed	thyme, rosemary, basil, oregano, vanilla, mace, clove, nutmeg
Vegetables	iceberg lettuce, asparagus, fennel, turnip, cherry tomato, bell pepper (raw, unpeeled), radish	red beet, mushroom
Techniques that enhance . . .	poaching, cooking	baking, grilling, (deep-) frying
Condiments	raw onion, caper, mustard, pickle	garlic, truffle

Complexity is not just a matter of adding more flavors. There are flavors that boost each other and others that degrade. This is one of the mysteries of gastronomy. In the process of composing a dish, chefs are well aware of this phenomenon. "Flavor feeling," knowing if something will fit or not, is often regarded as an innate talent. Great chefs know instinctively what they are missing or what should be added. Just putting more on a plate without a certain coherence often leads to "flavor noise," not to complexity, elegance, or refinement. Just like in a musical composition or a good painting, the tones, colors, and composition fit together harmoniously. Complexity at its best is a rare feature and makes things valuable or expensive—all the more reason to put complexity on the table only when the situation is right.

With complexity, all flavors may be intricately interwoven. This can mean that the harmony is not directly recognized. Therefore, complexity often requires experience. This explains why experts may assess flavor differently than amateurs and is also why the role of the sommelier, waiter, or advisor in a restaurant or shop may be more important than often thought. If necessary they can help the taster in recognizing the complexity and in doing so increase appreciation. We will discuss this interesting subject in Chapter 7. In this chapter, we remain focused on the theory.

BUILDING THE THEORETICAL MODEL

We now live in the twenty-first century. We have been to the moon and lived in space. At the European Organization for Nuclear Research, the world's largest particle physics laboratory, the basic building blocks of the universe are being studied. These are all highly sophisticated endeavors. And yet, flavor—something that each and every one of us confronts daily—is just starting as a science. Universal flavor factors can be used to compose dishes and to find the proper beverages to match them. But there is more. They also provide the fundamentals for building a theoretical model. You may not think so, but I find that fascinating. In science, having a model is a wonderful gadget. Models open academic doors, allow scrupulous testing and assessment, facilitate discussion, and provide a foundation for other people to build upon. Or, in the words of Albert Einstein: "it is the theory that decides what we can observe."[1]

Any theoretical model requires mathematical language. In this case, this is done by putting values on scales. The scales have been introduced, but we need to take flavor intensity and flavor type together and use flavor richness instead as an overall scale that integrates both the intensity and flavor type. Flavor is built upon mouthfeel and flavor richness.

Mouthfeel is presented in Figure 3.1, which shows four quadrants each with its own basic characteristics. They can be described as follows:

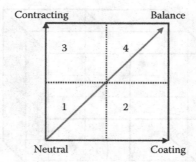

FIGURE 3.1   GRAPHIC PRESENTATION OF MOUTHFEEL.

1. Neither contracting nor coating but neutral, which can turn into dry (examples: rice, plain bread, mineral water)
2. Coating dominates over contracting (example: vegetable oil)
3. Contracting dominates over coating (example: vinegar)
4. Contracting and coating are both present and more or less in balance (example: vinaigrette)

The two dimensions of mouthfeel already present a good structure with ample possibilities for classification. When we add flavor richness to the model, the foundation for a three-dimensional structure is there, as Figure 3.2 shows.

This basic structure can be extrapolated into a three-dimensional model: the flavor styles cube (Figure 3.3). We now have eight subcategories—the flavor styles. Every style has specific characteristics that can be used to identify the location of a food or drink. Flavors that

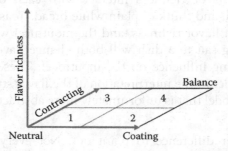

FIGURE 3.2   ADDING THE THIRD DIMENSION OF FLAVOR: FLAVOR RICHNESS.

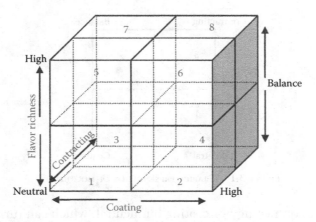

FIGURE 3.3   THE COMPLETE MODEL: THE FLAVOR STYLES CUBE.

can be identified close to one of the corners can be considered to be an icon of that flavor style (see Table 3.2).

This model is a depiction of the world of flavor and serves as an instrument for classification. It does not imply that there are only eight flavors. Compare this to light, for example. The flavor styles correspond to the colors of the rainbow. Every color has many shades, and they fade beautifully into each other. There is an obvious difference as well. The exact position of any color is precisely known. If you go to a paint shop or a printer, you can give a number to get the desired color. With flavor, we are as yet nowhere near such exactitude. It is even hard to conceive that we would ever get to such detail. We have, however, made a huge step forward from where we were.

Mouthfeel and flavor richness interact with each other. Take, for instance, neutral, and think of plain white bread. Toasting it will lead to an increase in flavor richness, and the mouthfeel will become dry. Likewise, adding salt to a dish will both change flavor richness and have a contracting influence on the mouthfeel. These are important aspects to consider in the interpretation of the flavor styles cube. It is a mathematical model that is unfortunately more abstract than the world of flavor deserves.

There is another difference to point out. Not every flavor style is equally important. Flavor, as we are concerned with, is mostly made by

TABLE 3.2
The Basic Characteristics of the Flavor Styles

Flavor Style	Primary Flavor Dimensions			
	Contracting	Coating	Flavor Richness	Description
1	Neutral	Neutral	Low	neutral, light
2	Low	High	Low	round, smooth, supple, creamy
3	High	Low	Low	fresh, sour, contracting
4	High	High	Low	can be eaten/drunk continually, simple
5	Dry	Dry	High	robust, solid, powerful
6	Low	High	High	full flavor, ripe flavor, filling
7	High	Low	High	pungent, spicy, hot, explosive
8	High	High	High	complex, differentiated, subtle

people—chefs, food technologists, winemakers, and so on. The market for popular flavor styles is therefore bigger than for the ones that are less useful. To illustrate: if you were a winemaker, would you make a wine that fits category 1? It would be neutral wine without a specific character and hard to sell, to say the least.

All flavors can be positioned within the cube. Foods and drinks are complex mixtures of different components, and the individual components influence each other. Changes in mixtures may lead to different profiles. This implies that the position of real food or drink in the flavor styles cube is dynamic. The corners represent clear flavors. As previously noted, products that are positioned close to a specific corner can be considered icons of that specific flavor style. Flavor styles within the cube that lie above each other are closely associated. These flavors are similar in mouthfeel but differ in flavor richness; their basic characteristics are discussed tentatively as follows.

DESCRIPTION OF THE FLAVOR STYLES

In this section, the flavor styles are discussed in detail. Table 3.3 gives an overview of practical examples of each style.

FLAVOR STYLES 1 AND 5, NEUTRAL AND ROBUST

Flavor style 1 is neutral. It represents tasteless or very little taste in mouthfeel and in flavor richness. Distilled water can be situated close to this corner. In the history of flavor classification, Fernel proposed the term "insipid."[2] This term still exists today, but it has a negative association. We prefer to use value-free terms as indicators of the specific flavor styles. Neutral would be an appropriate term to characterize this flavor style.

Flavor style 5—the extension of style 1—is robust. Mouthfeel is basically dry, and several influences may account for this dryness. The first is absorption of saliva as happens with dry solid foods such as toast, biscuits, crackers, and chips. Next, tannins are known to affect saliva, making the mouth rough.[3] Astringency is the appropriate term for this kind of bitterness; it is a trigeminal sensation rather than a gustatory stimulus (discussed in Chapter 2). Some young red wines can be "drying" or "puckering."[4] Likewise, some other bitter tasting substances such as strong coffee or tea and dark chocolate with a high percentage of cocoa can be astringent, bitter, or acrid.[5] The flavor qualities harsh and astringent that were originally included in the classification of flavor by Aristotle would fit here. Although the characteristics of products around

TABLE 3.3
The Flavor Styles and Typical Examples

Flavor Style	Typical Examples	
	Food	Beverages
1. Neutral	rice, white bread, zucchini, poached/steamed fish	plain water
2. Round	spaghetti, milk chocolate, banana, melon, avocado, young Gouda cheese, butter, plain oil	milk, wine with a touch of sweetness
3. Fresh	apple, orange, lemon, green salad, pickle, oyster, tomato, fresh goat cheese, vinegar	orange juice, acidic white wine such as Sancerre, Muscadet, Riesling
4. Balance low	vinaigrette, creamy tomato sauce	yogurt, balanced daily wine
5. Robust	dark chocolate, grilled red meat	espresso, strong tannic red wines with wood character
6. Full	vanilla, caramel, whipped cream, vegetable oil, peanut butter	sweet, concentrated wine, fortified wine
7. Pungent	peppermint, mustard, horseradish, red pepper, ginger, spicy dishes	some sparkling wine
8. Balance high	braised meat, mushroom, umami	rich, full-bodied yet balanced wine (not sweet)

the flavor style 5 corner may not always be instantly pleasant to consume, their role can be functional in relation to other food components.

FLAVOR STYLES 2 AND 6, ROUND AND FULL

Flavor style 2 is round, and its counterpart, flavor style 6, full. Both styles are characterized by the thin and coating layer that is left behind in the oral cavity by fats and/or dissolved sugars.[6] Fatty as a flavor characteristic was also introduced by Fernel, and sweet has always been recognized. However different, both elements share the characteristic of being coating in mouthfeel. Physiologically, these flavor aspects are sensed by the mechanoreceptors of the trigeminal system. Their acuity is quite remarkable: even small differences in viscosity can be perceived and particles as small as 5 µm (that is 0.005 mm, just slightly bigger than one bacterium) can be detected.[7] The difference between flavor style 2 and 6 is explained by flavor richness.

FLAVOR STYLES 3 AND 7, FRESH AND PUNGENT

Flavor style 3 is fresh. The flavors in this section are basically characterized by contracting elements in foods and drinks. Acids and salts are well-known substances that have a contracting influence. Acidity is also reported to increase saliva flow.[8] Physiologically, these substances are reported to enter the ion channel of taste cells directly.[9]

Flavor style 7 is pungent. It is here where we principally find the oral irritants, such as capsaicin, gingerone, and CO_2, which all activate trigeminal pain pathways. Those reactions are mediated by the nociceptors of the trigeminal system that also enter directly into ion channels.[10,11]

FLAVOR STYLES 4 AND 8, BALANCED, FROM LOW TO HIGH

Flavor styles 4 and 8 are both balanced, but the flavor richness is different. Flavor becomes balanced if coating and contracting forces more or less compensate each other. Furthermore, the so-called fifth basic taste, umami, is principally active in these flavor styles. Basically umami stands for the taste that is enticed by glutamic acid. It is widely used in the food industry to enhance flavor, palatability, and overall preference.[12] Statistical analysis of terms that are used to describe umami flavor showed a correlation with ratings for complexity, mildness, brothy, and meaty.[13] These descriptions of flavor perfectly suit these styles. Furthermore, umami is reported to be transduced by special G-protein-coupled receptors—different from the effect of sweet, fats, acids, and salt—which justifies its separate position.

HOW TO GENERATE A FLAVOR PROFILE

Everything that we eat and drink can be classified, and this is basically not very hard to do. Once you know what mouthfeel and flavor richness is about, recognizing the flavor components becomes like a second nature. The sum of the components is called the flavor profile. Spoken mathematically, the profile represents the position of a flavor in the cube. The tasting may take some practice and time to get the figures right. It becomes easier if two products can be compared. For instance, when two white wines are tasted next to each other, it is likely that the one differs from the other in mouthfeel, flavor intensity, and flavor type. To limit the chance of individual misinterpretations, results become more accurate if the tasting is done by a group of people.

It is important to make a distinction in the level of detail needed to score products. If the mission is to locate the position of a certain product in the flavor styles cube, it is sufficient to score on the three dimensions: contracting, coating, and flavor richness. If the mission is food and beverage matching, a greater level of detail is required within the framework. In the course of this book, it will become clear that we need to know about the type of coating or contracting, if sugar or acidity play a role, and we need a characterization of the flavor style. Therefore, a specific scoring form has been developed and is shown in Figure 3.4.

Each of the six axes is independent, although some of the axes seem to be each other's opposite. Every axis is scored on a scale from low/hardly present to high/very present. We will briefly introduce them to prevent misunderstandings.

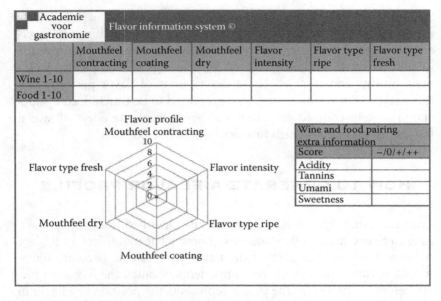

Academie voor gastronomie	Flavor information system ©					
	Mouthfeel contracting	Mouthfeel coating	Mouthfeel dry	Flavor intensity	Flavor type ripe	Flavor type fresh
Wine 1-10						
Food 1-10						

FIGURE 3.4 THE SCORING FORM FOR WINE AND FOOD MATCHING FROM THE DUTCH ACADEMY FOR GASTRONOMY.

- Flavor intensity: the volume of flavor
- Fresh: the aroma type
- Contracting: the type of mouthfeel of contraction by, for example, acidity, salt, and also spiciness
- Coating: the mouthfeel of fats that leave a layer behind
- Dry: the type of mouthfeel from absorption of saliva or changing the properties of it
- Ripe: the aroma type
- Acidity: the level of acidity
- Tannins: the dryness of some young red wines, green tea, and occasional whites
- Umami: which could also be called "mouthfullness." The type richness that influences mouthfeel that is not from fat or sugar
- Sweet: the level of sweetness

Scoring is done by taking a sip or a bite of the product and chewing or leaving it in the mouth for five seconds before swallowing. A mark is

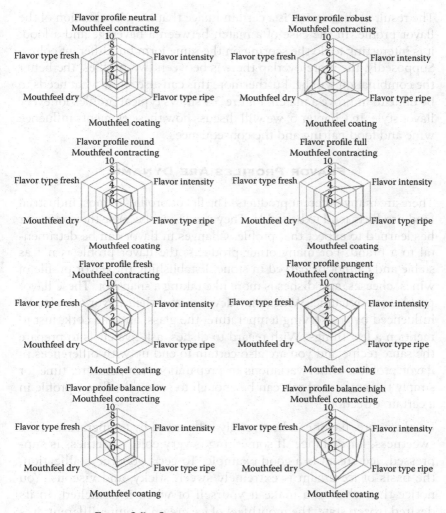

FIGURE 3.5   SOME TYPICAL EXAMPLES OF FLAVOR STYLES.

then given for each of the elements. Taking this time before scoring is essential, especially with regard to mouthfeel. Wines, for instance, all have a degree of acidity; therefore, most will end up contracting if you wait too long. Furthermore, substances in the mouth are mixed with saliva, so mouthfeel will change in all cases.

The result of the scoring is a certain image that is a visualization of the flavor profile. In the case of a match between a beverage and a food, it is interesting to do the scoring on the same form but different colors. Supposedly, the more overlap there is between two profiles, the better the combination will be. Furthermore this can also show what needs to be done to improve a match. Figure 3.5 shows typical profiles for every flavor style. In Chapter 6, we will discuss how these elements influence wine and food pairing and the consequences.

FLAVOR PROFILES ARE DYNAMIC

There are many different products. The flavor profiles of most industrial food products tend to be stable. They need to be because the consumer has learned to expect that profile. Changes in flavor can be detrimental to a brand. For many other products, the flavor profile is not as stable and cannot be carved in stone. Establishing the flavor profile of wines, cheeses, and dishes is more like taking a snapshot. These flavor profiles are dynamic. Wines may develop, and their flavor can also be influenced by the serving temperature, the glass, and the cork, just to mention a few things. With regard to dishes, ask five chefs to prepare the same recipe and you are also certain to end up with differences in flavor profiles. Minor variations in preparation, temperature, time, or simply the amount of salt can be enough to swing the flavor profile in a certain direction.

Furthermore, flavor can be strongly influenced by temperature. Take sweetness, for instance. If something is very cold, sweetness is suppressed. Ice cream is a good example. To get a good crystallization, the basis of ice cream is extremely sweet, sticky, and viscous. You notice this when you make it yourself or when it is melted. In its desired frozen state, the mouthfeel of ice cream is quite different; it is much more pleasant, mostly refreshing, and certainly not as coating. For the same reason, the serving temperature of wines can influence their flavor, especially if the wines have some residual sugar. If they are served on the cool side (<10°C) the balance moves toward contracting because the sweetness is suppressed. The acidity shows more strongly. Serve it a few degrees warmer and the mouthfeel moves toward coating. It is an interesting experience. Even more so when this same wine at the lower temperature gives a better match than slightly warmer—or the other way around. A sommelier in a restaurant can actually improve a match just by playing with the temperature of a wine.

Balance in mouthfeel can also be altered by a change in dryness. Using French fries as an example, the crust needs to be crisp. If you leave them on the table for a period of time, they will quickly lose their crispness and their appeal. For the same reason, bread needs to be freshly toasted. There are many similar examples. In these cases, the crispness is even essential for liking the product.

Besides mouthfeel, flavor richness is also influenced by temperature. Many aroma compounds need a certain temperature to become volatile. Anything that is served too cold will appear to be less rich in flavor than when served at the proper temperature. The word *proper* should be used in regard to specific categories. The correct temperature of, for instance, coffee and tea is different than that of wine and beer or of cheese. The first two are best served hot, while the temperature range for wines is somewhere between 10°C and 18°C. Most cheeses are best when served at room temperature. When chilled, the flavor of a creamy cheese such as Brie or Camembert is not as rich as when the same cheese is savored at room temperature (around 22°C).

The flavor of many cheeses is also strongly influenced by aging, just as are many other products. In a natural maturation process, enzymes work on proteins. Such a process may take a year or more. Think of how the flavor of cheeses such as Gouda or cheddar develops. In some cases, mere weeks of aging can cause a big difference in flavor. Beverages such as wine, whiskey, or brandy can also gain flavor richness in aging. This may be due to complex interactions between aromas and alcohol or evaporation. In general, flavor richness tends to increase, and mouthfeel changes as well.

If all such details are as important as that, it becomes clear that general suggestions such as white wine with fish, red wine with meat, and so forth are rather silly. Mostly, it is not the main ingredient that determines the good combination. Preparation, garnishes, and details such as serving temperature are often more important. In other words, to know more about its flavor profile, you need to know if the fish is fried, poached, or grilled; served with a vinaigrette, mayonnaise, or a cream sauce; and with rice or French fries—just to mention a few examples. The same is true for wines. The quality of the grapes and the methods used by the winemaker determine the flavor profile. This explains how wines from the same region and made from the same grape variety can still be very different. The label does not tell the difference. You need to taste the wines to know.

THE ROLE OF THE BASIC FLAVORS

The flavor styles model provides an understandable depiction of the world of flavor that is rather easy to use. How does this model relate to the basic flavors with which we have grown so familiar? Obviously they still exist, but they are considered with regard to the influence they have on mouthfeel and flavor richness. Acidity, bitter, and salt have a contracting influence; sweetness is coating. Umami as well as salt have a big impact on the flavor richness. Let us review each of them in more detail.

SWEET

It is generally presumed that we like sweetness right from birth. The flavor is of course directly associated with sugar: water-soluble crystalline carbohydrates that are mostly obtained from sugar beets and sugarcane. Most common are the so-called monosaccharides: glucose and fructose, and disaccharides: sucrose (a mix of glucose and fructose), lactose, and maltose. There are also more complex polysaccharides such as pectins, gums, starches, and cellulose, which are discussed in the section "Carbohydrates" in this chapter.

Sugars are widely used in flavoring and not only because people like sweetness. Sugars have other qualities as well: they are natural preservatives, and they contribute to coating in mouthfeel, or fullness. In the flavor type, sugars will generally lead to "ripe." Furthermore, sugars have the capacity to mask bitterness and acidity. Those are attractive characteristics and, therefore, sugars are extensively used in the food industry. Many of the popular food products that are consumed on a large scale, including the ones that do not taste sweet, contain high fructose corn syrup (HFCS). HFCS is retrieved from the starch of corn (maize), and its extensive use has resulted in debate and study. Researchers point at the difference in metabolism of glucose and fructose. Glucose functions as sugars are expected to work: they give energy and are digested by the secretion of insulin. Another hormone, leptin, is produced as well that gives a feeling of satiety, which should help in preventing eating too much. Fructose, on the other hand, seems to function quite differently. Instead of giving energy, it is metabolized similarly to alcohol and is directly related to forming fat. This may play a role in the growth of obesity in the Western world.[14]

Sugar is not only industry's favorite, modern young cooks tend to use more sugar in the kitchen than in the past. In mayonnaise and salad dressing, for instance, it has almost become a standard ingredient. Without going into the more philosophical discussion as to whether this is a good development or not, it should be noted that the use of sugar in dishes is not without consequences for their combination with wines. Wines that are completely dry (meaning that they do not contain any residual sugar) will appear to be more acidic, tart, or sharp next to a dish with sugar. This happens even when just a small amount of sugar is used. The combination rule is, therefore, that the grams of sugar in the wine and the dish should be more or less equivalent. Interestingly, vegetables such as beets, carrots, bell peppers, and onions can acquire a certain sweetness in preparation, mostly in the oven and after quite some time. This type of natural sweetness does not interfere with dry wines.

From a nutritional perspective, sugars are an important nutrient. Sweetness is a source of pleasure; it is meant to be a treat and not to be consumed in large quantities. Yet, for many people, this is the case. The food industry has come to rescue with "light" products: sweet tasting food additives containing little or no calories. They contain substances such as aspartame or saccharin. Although there are natural and synthetic sweeteners, the majority of what is used in industry is artificial. We will not go into a discussion here about the health risks that may be involved. From a flavor point of view, it is important to note that these products mainly elicit sweet taste but lack the other properties of sugar. They do not caramelize and do not give fullness in mouthfeel.

Replacing sugar with sweeteners in a recipe will require other changes as well. In the food industry, so-called bulking agents are added that give fullness; an example is maltodextrin.

Something else worth noting is the effect that labeling has on consumption. When something is called light, people tend to consume more because they think that there are no calories. These days, many young people drink a soda when they are thirsty, instead of water. Such profound changes in our dietary pattern need to be taken into account in an effort to understand and ultimately solve the increasing health problems that we face (obesity, cancer, food allergies, etc.).

SALT

Salt is a mineral that can either be obtained from seawater or rock deposits. It is by far the oldest of food seasonings, and it is used all over the world. It is a flavor enhancer, which means that many substances taste better with a pinch of salt. Therefore, it has an important influence on flavor richness; in flavor type, salt is predominantly fresh, and bitter is slightly suppressed by salt.

Salt is more than just a seasoning. It is an essential nutrient that fulfills many important roles in the body. Humans and animals exert themselves considerably to obtain the salt they need. Furthermore, it is an indispensable ingredient in a wide range of preparations that include the making of cheese, bread, and all kinds of foods in general. For many thousands of years, salt has been used to preserve food, and especially meat. It has always been important, and that importance is hard to overestimate. Roman soldiers received a *salarium* for their services: the amount of money needed to buy salt. Think of that when you receive a salary.

The chemical compound for salt is $NaCl$. Its major components, sodium and chloride, are essential for all living creatures in order to survive. Salt regulates the water balance of the body, and the nervous system uses the sodium ion in the transmission of electrical signals. If you are not well informed about food and believe everything that is said, this may surprise you. Isn't salt something that is very bad for you? Do we not need to reduce the consumption of salt wherever we can? Apparently salt cannot be all that bad because its use is even indispensable in many respects. Nonetheless, it is good to be aware of overconsumption. Sensorially, it is easy to become accustomed to a certain level of salt, and this may lead to a gradual rise in use. If this occurs, it is sensible to cut down on salt consumption. This may take some time and, in the beginning, foods will seem to be too low in flavor richness, but that will gradually improve.

Besides the salt that we know, there are other substances with a salty flavor—potassium chloride (KCl) for instance. It is not as salty as NaCl and gives a bitter aftertaste. If it is used in the food industry in an effort to reduce sodium, other substances are needed to cover up such undesired effects. The one challenge leads to another and keeps research departments with food scientists busy. Professional chefs and people who love to cook at home just reduce the amount of salt and use some herbs, concentrates, or acids to solve the problem.

In addition to plain table salt, there are all kinds of fancy salts, sea salt, special rock salts, and so on. Some people argue that this is all nonsense; we should use NaCl and nothing more. For those people, it is good to note that many minerals may be bound to the salt. Compare it to water. Although all waters are nothing more than H_2O, put the labels of some mineral waters side by side and you will see big differences. Taste the waters and they will not all be the same. Truly there is a difference in salts, and arguably the most elegant of all salts is the hand-harvested sea salt, *fleur de sel*.

SOUR

Acidity will not rank as high in a list of the most popular flavors as sweet and salt, and yet it is an important substance in flavor. Acidic products are mostly refreshing and often considered to be light and lively. In mouthfeel, acidity is on the contracting side, and it stimulates saliva flow. Just like salt, acids can be used as preservatives. Pickles are the most common example.

Natural fruits and vegetables contain acids, but there are different kinds. Citric acid is present in oranges, lemons, limes, and other types of citrus fruits. Malic acid is typically found in green apples but also in many other fruits. Lactic acid is associated with milky flavors and is predominant in yogurt and sauerkraut. Oxalic acid is found in rhubarb and, to a lesser extent, in tomatoes and spinach. A special mention goes to ascorbic acid (better known as vitamin C), which is not only required by our body, it is also used as an antioxidant in food products and wine. Carbonic acid—that gives the fizz to beverages and hardly tastes sour—also needs to be mentioned.

The list of acids is much longer, but the ones discussed here are important in the world of flavor. If you go through the examples, you realize that not all acids taste the same. Aromatically they are different, and some are more acidic than others. If you apply this to the flavor styles cube, this would be the flavor richness of acids. Lactic acid is not as intense as citric or oxalic acid. Add salt and/or pepper to it and it will become even higher in flavor richness, more contracting, and, ultimately, it will be pungent.

Wine grapes contain primarily tartaric and malic and to a lesser extent citric acid. The content of malic acid is directly associated with the grape variety and the climate. In warmer regions or years, there is less malic acid. In the winemaking process, malic acid can be broken down to lactic acid, which is softer. This malolactic fermentation, as it is called, can be used as a tool to control the acidity of the wine. When wine gets too old or oxidized it can turn into vinegar, especially when enough oxygen is available. When this happens, the wine may have lost its role at the table but has acquired a new function. Vinegar—or sourness in general—is an underestimated part of flavor; many dishes and flavors get better by adding some acidity.

Acids add up. This is important to know with regard to wine and food pairing. All wines have a certain degree of acidity and, for a good match, the dish needs this as well. However, not too much—if the dish is already rather sour, a wine with high acidity will be perceived as even more acidic. One last remark: we quickly adapt to sourness. The first sip of a wine or a bite of an apple is likely to be considered as too sour, but after that it gets better. Never judge the acidity of something too quickly.

BITTER

Bitter is likely to be found at the bottom of our popularity table. Few people will say that they love bitter. For one thing, it is an acquired taste (meaning that you have to learn to appreciate it). The explanation for this phenomenon that people seem to have accepted is that many toxins have a bitter flavor. This leads to the suggestion that we need to be reprogrammed to learn that bitter may be not so bad after all. Is this true? If you look at the most popular products that are universally consumed on a large scale, you get another picture. All over the world, coffee and/or tea, beer and/or wine, and chocolate are favorites. The common denominator of these products is bitter. And it gets even more interesting if you take into account what the true aficionado of these products thinks of them: the more bitter the better; this is true all along the line whether it is coffee, beer, or chocolate. I call it the bitter myth. We actually seem to like at least some bitters, although generally not too much. This is not surprising because too sweet, too acidic, or too salty are also not liked. In the kitchen, however, bitter flavors are sought after; they give dishes character and depth. Bitter tastes in just the right proportion contribute to harmony and balance.

This makes us wonder why we developed such bad feelings about bitter. The answer is found in science: there are problems on the receptor side and on the tastant side. Recent studies show that humans are very sensitive to tasting bitter substances. It appears that we have many different taste receptors to identify all kinds of bitter compounds.[15,16] This is likely the result of a defense mechanism because many toxins taste bitter. Furthermore, bitter is not directly associated with a specific chemical or chemical group. A wide range of compounds can elicit bitter taste and not all bitters taste the same. However, one group does stand out: alkaloids. This is a large group of substances that are found in plants; they mostly contain nitrogen. Indeed, they are often toxic to certain organisms, but quite a few of them have beneficial qualities as well. Some examples are caffeine, nicotine, cocaine, morphine, and quinine. In order to study bitter and obtain comparable results, it is best to select one bitter. Generally, quinine is used in the practice of taste research.

Quinine is a drug that comes from chinchona tree bark and is very effective in treating malaria. It reduces fever, is anti-inflammatory, and is a painkiller. It is also bad tasting because of its extreme bitterness. Without exception, everybody dislikes the flavor of quinine, which is even seen as an advantage in this case because drugs are not meant to be consumed like candy. The real problem is that it has "always" been used in scientific flavor research. This choice has had a severe negative impact on the reputation of bitter as a basic taste. It is not bitter that we dislike, it is quinine in particular. Next to coffee, beer, wine, and chocolate, most of us like lettuce, chicory, and endives—especially if well prepared. Most of these substances have beneficial qualities. Take vegetables, for example. The *American Journal of Clinical Nutrition*[17] points out that most vegetables contain antioxidant nutrients, and most of these antioxidant nutrients are bitter. In gastronomy, bitterness is countered in preparation by adding sauces, sugar, or some fat.

In terms of the universal flavor factors, bitter is on the contracting side, and in flavor type it is primarily ripe. Furthermore, it seems that flavor intensity and bitter flavors may be related. An example would be the role of certain techniques in the preparation of food. The process of baking, grilling, or frying raises the flavor intensity and, simultaneously, bitter tones will emerge (of the type that are generally considered to be pleasant, by the way).

A special characteristic of some bitters must be mentioned as well: astringency, which is bitter and dry. Tannins are a well-known example. The bitter of cacao and coffee can also be classified as drying. This happens when certain compounds bind the proteins of saliva, which makes that saliva lose its smoothness; it becomes dry. Not all bitters are astringent, however.

Umami

Umami is the new kid on the block, even though its existence has been known for more than a century. During this period, many misunderstandings have arisen, such as it is a flavor of Asian cuisine; it is not a flavor but only a flavor enhancer; and it is the same as monosodium glutamate (MSG)—artificial and something to be avoided.

To start with, umami is definitely not an invention of Asian cuisine (although the word *umami* means "delicious" in Japanese). It has been known and appreciated for centuries. It gives mouthfullness and roundness, and is also described as balancing, blending, and as giving continuity, thickness, and a general feeling of satisfaction. These effects on flavor are attributed to glutamic acid, one of the most abundant of all amino acids found in food protein.[18]

An ingredient called *garum* or *liquamen* appears in most recipes of the oldest cookbook that has survived, *De re coquinaria*, by the Roman gourmet Apicius (first century AD). It was a fermented fish sauce that was widely used as a condiment to enhance the flavor of various dishes. Originally, it came from the Greeks. A similar sauce is still widely produced today all over Southeast Asia. The main component responsible for flavor enrichment is an amino acid called glutamic acid, which was discovered and patented by the Japanese professor Kikunae Ikeda in 1908. Initially, he proposed calling it glutamic taste. Ikeda founded the company that commercialized umami (Ajinomoto, which means "the essence of taste"). Other umami condiments that are widely used include the various soy sauces (Kikkoman, Ketjap) and Ve-Tsin.

This may all sound closely associated to the Asian culinary practice after all. But the list of umami usage is much longer and extends to other cuisines as well—for example, in Italian concentrated tomato sauce, American ketchup, and British Marmite and Worcestershire sauce. The concentrated stock called *glace de viande* that is the basis of many French recipes is also rich in free glutamic acid. Cuisines all over the world have a natural umami component of some sort. To complete the list, we could add Parmesan cheese, traditional Dutch Gouda, Roquefort, cured ham such as the Spanish Jamon Iberico, and dried mushrooms. Last, but in no way least, human breast milk is rich in umami.

Besides glutamate there are three kinds of 5'-ribonucleotides that contribute to the umami sensation: inosinic acid (IMP), found in animal sources; guanylic acid (GMP), more abundant in plant based foods and mushrooms; and adenylic acid (AMP), which is present in fish and shellfish. There is a strong synergy between glutamate and these ribonucleotides, implying that the umami taste is strongly enhanced by a combination of the two.[19,20]

Then there is the question of whether umami is a flavor enhancer or a singular flavor as well. This misunderstanding can be answered by looking at the definition of basic tastes. A substance is called a basic taste if the sensory signal is registered by papillae on the tongue. Clearly this is the case with umami and, therefore, a nonissue. In fact, sweetness, saltiness, and acidity enhance flavor in their own way. One caveat to note, however, is that umami lacks a specific flavor and is not as easy to recognize as others. To me, it is coating salt. Normal salts are contracting, but umami seems to interact with saliva, making it fuller. Umami makes other flavors more interesting. It is like a layer of varnish on wood: the wood gains depth and shines. And just like layers of varnish, a few layers suffice; only small quantities of umami are needed to have an effect.

The sodium salt of glutamic acid is called monosodium glutamate or MSG. Several varieties are massively used in industry and go by the E-numbers 620, 621, 622, 623, 624, and 625. Ve-Tsin is popular in Chinese cuisine, but not all chefs know that just a little bit is enough. In some cookbooks, you see tablespoons of Ve-Tsin used. And although agencies such as the U.S. Food and Drug Administration generally classify MSG as safe, it is very possible that an overdose might have an adverse effect on humans. This effect has been referred to as "Chinese restaurant syndrome" and may result in warmth, headache, chest pain, rapid heartbeat, and a heavy feeling. Glutamate is a one of the nonessential amino acids, which means that it can be produced from other amino acids by the human body itself. The extra glutamate that is consumed is absorbed quickly and could spike blood plasma levels, which might explain the problem.[21] Examples such as this suggest we should be vigilant about the use of glutamate in food and especially the types that are used in industry. Today, "'yeast extract" is seen more often on labels and avoids the use of an E-number. It is surprising to see advertisements for "E-number free, new recipe." Technically that may be correct, but the practice is hardly acceptable.

Together with aspartame, glutamate belongs to the class of excitotoxins. High levels of these toxins may cause damage to parts of the brain and give rise to the development of a variety of chronic diseases. Although reports are somewhat contradictory about the effects, there are reasons for concern. The relationship between MSG and umami is similar to that of aspartame and sweetness. If there were reasons to be suspicious about aspartame, it would be silly to condemn all compounds that elicit sweetness. Likewise, umami should not be condemned if MSG is suspected. There are many natural products and well-known fermented products that elicit the umami effect, many of which have been used for ages. One never hears of problems concerning the consumption of traditional products that are rich in umami such as Parmesan cheese, tomato sauce, or Iberico ham.

THE BASIC FLAVORS FROM A DIFFERENT PERSPECTIVE

To conclude this section, let us add another perspective to the basic tastes. Salt and sour have a similarity. Both work through ionic pathways—the signals go right through taste cells and are not registered by receptor cells on the outside. In flavor, they give what can be called vibrancy, an aspect of liveliness, a sparkle. Minerals such as CO_2, pepper, ginger, and other spicy elements have a similar role in flavor. Unlike sweet and umami, which seem to add more to the structure of flavor, vibrant flavors fall into a category with, for instance, fats or carbohydrates.

Vibrant ingredients add dynamism and energy to dishes. They can hardly be missed. Chefs need to understand the role of these ingredients and see to it that they are included.

THE ROLE OF THE BIG FOUR
BASIC FOOD MOLECULES

To conclude this chapter, we will discuss the role of the four basic food molecules: water, fats, carbohydrates, and proteins. Each of these elements has properties that are interesting to know about in order to appreciate their role in the flavor profile and ultimately in the preparation of foods.

WATER/MOISTURE

Water is essential for life. It is the most important component of nearly all foods and—obviously—beverages. By weight, raw meat is about 75% water; in fruits and vegetables this may be as high as 95%. This means that a cucumber, for instance, contains more water than many mineral waters. Whatever their disguise, fresh foods can be seen as packaged water plus substances such as proteins, fats, carbohydrates, vitamins, and minerals. Although it is the simplest of food molecules, the significance of water is spectacular. It contains just three atoms—two hydrogen and one oxygen, no calories, and it is most effective in quenching thirst, which makes one wonder why so many people today turn to sodas when they need something to drink.

The principal properties of water are well known: it boils at 100°C, and it freezes at 0°C (both measured at sea level). The moisture content in foods influences mouthfeel; it gives juiciness. Maintaining the moisture content is one of the major challenges for preparing foods at higher temperatures. The opposite of heating is freezing. We have seen before that (serving) temperature influences mouthfeel. A low temperature makes the mouth contract. But that is not the only influence. Crystallization is also important. Big crystals are chunky and rough. In the best ice creams, the mixture is kept in motion while being frozen. Given that the mixture has the right quantity of sugar and/or fat, this results in small crystals and gives the ice cream its desired smooth texture.

Freezing is also an important technique in conservation. Not only does it stop the growth of bacteria, but there is another effect as well. Water molecules are bound in cell structures. Water expands when frozen. This implies that the cell structures will get damaged and may lose their water holding capacity. When thawed, the water runs out, often with detrimental effects to the flavor. The faster the freezing process takes place and the longer frozen substances have time to thaw, the less damage there is. The same happens in cooking; in the process of heating, cell structures may also get damaged, with the same negative effect. In general, retaining moisture must be considered one of the principal challenges of food preparation. We will discuss it in depth in the next chapter.

One of the most important properties of water is its polarity. Every water molecule has a positive and a negative end. Hydrogen atoms are positively charged. The negative oxygen feels attracted to the hydrogen on other molecules. Similar to a magnet, water molecules pull together and form what are called hydrogen bonds. This natural tendency is the basis of some important processes in the kitchen—very likely without the average person being aware of this.

The polarity of water also explains its effectiveness as a solvent for many substances: hydrogen bonds are not only formed among water molecules but also with other substances that have a certain polarity such as carbohydrates, alcohols, acids, and proteins. These molecules are larger and more complex than water and all have polar regions. The water molecules unite around them and separate them from each other. As such, parts of these other substances associate themselves with the water; they are dissolved. The polar regions of these molecules that associate with water are called *hydrophilic* (water loving).

Water is exceptionally important as a solvent. Many flavor components are dissolved in water. Every bouillon or stock is a wonderful illustration of this capacity. Starting with plain water and then simmering, flavor components sort out of their original structures (leaves, herbs, vegetables, etc.) and become part of the water.

LIPIDS: FATS AND OILS

Lipids comprise a large group of molecules that include waxes, sterols, fat-soluble vitamins (A, D, E, and K), and fats and oils, of course. In fact, fats belong to a specific subgroup of lipids: the triglycerides. Technically, they are a combination of a glycerol and three fatty acids. There are all sorts of fatty acids, and different compositions lead to specific structures and characteristics. Some differences are visible: those that are liquid at room temperature are called oils; fats are the ones that are solid. The explanation for this phenomenon is found in what is called the level saturation of fat. Saturated fats have a regular, orderly, solid structure. Vegetable fats are about 85% unsaturated and remain liquid.

Aromatic compounds are either dissolvable in water or in fat. Water and fats are in many ways the opposite of each other, but both are important in all kinds of foods and beverages. One reason has just been mentioned: there is a category of aromatics that is dissolved in fat. Take an orange, for instance: the juice has a characteristic orange aroma. The skin contains essential oils that have a slightly different character; nonetheless, they are also characteristic. Therefore, there is not one orange flavor. The part of the skin can be used in fatty substances, where they are soluble. In a watery solution, they would just float to the surface and not give much flavor. The other part of the orange needs a watery environment in order to be tasted. As discussed earlier, everything that is soluble in water is called hydrophilic. Lipids, including the aromas and other substances that are dissolved in them, are called *hydrophobic* (they fear water).

In normal circumstances, fat and water do not combine together. However, if they are blended with an emulsifier, it is possible to get a stable mix. This is the secret behind well-known foods such as mayonnaise and milk. Examples of emulsifiers are proteins and, in particular, egg yolk (the real acting agent is a diglyceride called lecithin), honey, and mustard. When a mixture is shaken, whisked, centrifuged, or otherwise processed, emulsifiers shield the lipid droplets from each other and form a blend that is more or less stable (at least for some time, depending on the mixture).

The role of lipids in food is particularly important, if only because of the general misconception that fats are bad and should be avoided. This never should have happened and, therefore, we approach fat from the positive side. We start with the quality mentioned earlier: fat is flavor. There is quite a category of aromas that are dissolved in fat. A simple experiment can illustrate this. Take lean minced beef and divide it into three portions. Add 10% pork fat to one portion, 10% lamb fat to another, and 10% of beef fat to the last. Fry these as a normal hamburger and ask people to guess the origin of the meat. It is highly likely that only the one with beef fat will be assessed correctly. The others will probably be identified as a hamburger of pork and lamb, although 90% is beef. Apparently 10% fat from these animals is enough to give the meat its flavor. This is just one illustration of the importance of fat as a carrier of flavor. Therefore, reducing fat presents a huge challenge: how can one compensate for the loss of flavor?

Looking at the aforementioned example, fat has another function as well—the hamburger with fat will definitely be juicier. Lipids give foods their roundness and fullness, although not exclusively. Carbohydrates and proteins can do this as well, as we have already partly discussed with regard to the use of sugars in foods. Fats and oils clearly contribute to a coating mouthfeel. Consequently, the low-fat challenge is even bigger: how can the desired mouthfeel be replenished? Our example implies that, to produce low-fat, light products, other substances must be added. These substances are not part of the original food and often are artificially made (read more about this in the Appendix to this chapter). Milk is in itself a wonderful illustration. Let raw milk stand for a while and the fat (cream) floats to the top. Skim it off and you have low-fat, skimmed milk that is white, watery, and has little or no flavor. However, the low-fat milk you buy in the supermarket has a fuller taste. This can only be the result of additives, but these need not to be mentioned on the package. In fact, it is very hard to find out what has been done.

Fat is an important component of food that affects palatability and nutritional value. Palatability is enhanced in two ways: in mouthfeel and in flavor richness. Both are relevant to appetite and the pleasantness of food. Foods that are high in fat are often overconsumed, and many individuals find it difficult to reduce their fat intake. Consequently, there is a growing demand for fat replacements in foods. A prerequisite in the development of these replacements is an understanding of how fats are detected. Clearly, tactile aspects such as moistness, creaminess, and viscosity are important, but fatty acids also modulate ion channels of gustatory receptor cells. Glutamic and tannic acid are also registered by more than one sensory system (i.e., the tactile and gustatory systems). Still, creaminess is more than viscosity. Fat replacers that match viscosity do not capture the full sensory appeal of fat.[22]

A preference for low-fat foods may be triggered by fat's unhealthy image. Fat people seem unhealthy, and obesity is a serious and growing problem. Dietary studies, such as those conducted under the supervision of Harvard's School of Public Health, indicate that fat consumption appears to have little to do with making people fat or even ill.[23] Nevertheless, there is a real health problem related to high levels of blood cholesterol. This may damage arteries and, in turn, may lead to heart disease. These high cholesterol levels are strongly correlated with the consumption of saturated animal fats, and trans fats. Trans fats do not occur in nature but are the result of an industrial process called hydrogenation that changes the structure of unsaturated fats. They become solid (such as all margarines). It has been well established that consumption of these industrial trans fats is a risk factor for the development of heart problems. Although there is reason to be aware of the negative side of some fats, it is rather unfair to blame all fats for making people obese. In addition, the role of refined carbohydrates in relation to obesity should also be taken into account.

This seems even more unreasonable if you take into account that quite a few fats, and especially oils, are particularly good for the body. As stated before, lipids are a source of some vitamins. A low-fat diet will almost certainly lead to deficiencies. Furthermore, some fatty acids are especially beneficial, such as omega-3 and omega-6. These essential fatty acids are indispensable to the proper functioning of the human organism. Because humans are unable to produce them on their own, these fatty acids must be present in food. In this respect, they are much like vitamins, although these fatty acids must be taken in greater quantities than vitamins. Omega-3 fatty acids are found in fish oils of predatory fish such as tuna, salmon, swordfish, and mackerel. Omega-6 is known as linoleic acid and is primarily found in most vegetable oils. A large number of studies indicate health benefits, many of which can easily be found.

To finish our support of fat, it is wise to pay attention to the high caloric content of fatty acids. They deliver about 9 kcal/g. Carbohydrates and proteins are the two other sources of energy for the human body. Both provide a maximum of 4 kcal/g. Limiting fat in the diet means that the body needs to get its energy from these other sources. Of the two, proteins are more expensive and complex, while carbohydrates are less expensive and rather simple to use. A low-fat diet almost automatically leads to one that is rich in carbohydrates. And that is not all. Both fats and proteins are more satiating than carbohydrates. In other words, consuming lipids or proteins will make you feel you have had enough more quickly. If you are in the food production industry, which one would you prefer to use: lipids, proteins, or carbohydrates?

CARBOHYDRATES

Carbohydrates are a large group of molecules that are produced by all plants and animals in order to store energy. However, carbohydrates are not essential nutrients for humans. In other words, our system needs water, fats, and proteins but can function without carbohydrates. On the other hand, a gastronomic world without bread, pasta, cereals, jams, and desserts would have lost quite a lot of its appeal.

The scientific synonym for carbohydrate is saccharide. Different types include the previously mentioned mono- and disaccharides, which are smaller and generally called sugars; they taste sweet. In this section, we focus on the larger polysaccharides, which do not taste sweet because they are too large to trigger the appropriate receptor. This becomes clear if you look at the polysaccharide that is by far the most important in the kitchen: starch. It contains thousands of glucose units, but it is rather tasteless. However, it has other qualities.

The principle characteristic of starch is that it can give a smooth and coating mouthfeel. It is a wonder to observe in food preparation how starches such as flour, corn starch, or tapioca can thicken substances. This will be discussed in greater detail in Chapter 7. For now, it is important to note that starch leads to coating in mouthfeel. Also, it is hydrophilic or water friendly—a quality that is important with respect to its combination with beverages that are mainly composed of water. They are likely to blend more easily with starchy structures than with fatty ones, which—as we have noted—are hydrophobic.

In dry structures such as crackers or toast, starch is not particularly coating. It absorbs saliva and may turn the mouth dry. In combination with other foods, absorption may be a useful feature. Smoked salmon is rather fatty; together with toast, the fats are less pronounced. Rice, pasta, or potatoes are often used to produce this effect.

Starch is consumed on a large scale. Many of the so-called staple foods that universally constitute a major component of diet contain large amounts of carbohydrates. Think of the cereals rice, wheat, and corn; root vegetables such as potatoes and cassava; beans such as lentils and different peas; and fruits such as bananas and water chestnuts. In fact, everywhere in the world there is some kind of food eaten abundantly that contains a lot of starch. Furthermore, it is the major component of popular prepared foods such as bread, pancakes, pasta, and noodles. Beverages such as beer, vodka, or whiskey could not be made without starch. And if this is not already enough, many modern processed foods contain hidden carbohydrates such as sugars (including the omnipresent high fructose corn syrup), derivatives such as maltodextrin, and many modified starches.

Starch is not the only polysaccharide that is used in the kitchen. There are also gelling agents such as agar agar, gum Arabic, guar gum, alginates, gellan, and others. Some of those have come into fashion in recent years with the innovative "molecular" cuisine.

Finally, there are two polysaccharides present in our food in large quantities that determine structure and mouthfeel to a great extent: cellulose and pectin. They cannot be digested; they fulfill an important role in the functioning of our digestive track, giving the food mass and stimulating the functioning of the intestines. Often, they are designated as fiber. In ripening fruit, the exact state of the cell walls determines the mouthfeel: when an apple, pear, or peach is mealy, it is beyond the proper degree of ripeness. At that point, the connections between the cell walls, provided by pectin, are destroyed and even the cell wall structure changes.

PROTEINS

We conclude our discussion of the big four with the most challenging and complicated basic food molecules: proteins. The properties of the other food molecules are rather uniform and so is their role in the flavor styles model. This is not the case with proteins. Proteins, such as a hair or a fingernail, are hard and stay that way. For us, these proteins are inedible (only specialized insects can cope with them). Compare that to what happens to an egg at different temperatures:

- At 55°C, there are no textural changes yet; an egg is still very liquid.
- At 60°C, the egg whites start to thicken or to coagulate (as the process is technically called).
- At 62°C, this also starts to happen to the egg yolk.
- At 65°C, you have a very soft-boiled egg; the whites are still somewhat loose and the yolk is still running.
- At 68°C, the yolk starts to set.
- At 70°C, you have what most people would call a medium-boiled egg—the whites and the yolk are set, but they are not dry yet.
- Around 80°C, the egg is called hard-boiled and is rather dry. Above this temperature, the yolk will begin to turn green.

In boiling water, these profound and irreversible changes happen within minutes; mouthfeel changes from very coating to dry. Not only raising the temperature can have such dramatic consequences. Acid, salt, or movement (whisking) may also change the structure of proteins.

Proteins are, together with fats, the main structural elements of living cells. Other proteins perform an almost infinite number of tasks within the cells and throughout the body by changing the structure of molecules. Proteins that perform these tasks are called enzymes. Just like starch, proteins are essentially long chains (polymers) of smaller molecules. The big difference with starch is that its chain is built up from just one type of smaller molecule. In the case of proteins, there are 22 types, called amino acids. A protein could comprise some 50 to many thousands of them.

From a gastronomic perspective, the breakdown of proteins is important. Enzymes are needed to cut protein chains into smaller parts. This breaking down of proteins, or rather the work of enzymes, is an important element of food preparation, just as in our body a similar action frees the amino acids that the body requires. Heating, especially in the presence of acids, is another method by which proteins can be broken down into amino acids.

In total, 22 amino acids are known to make up proteins. Animals in general (and therefore, humans) are able to compose their own proteins from amino acids taken up from food. But they are unable to produce all kinds of amino acids themselves: they must have a sufficient daily intake of amino acids, mostly in the form of plant and animal proteins. Ultimately, all amino acids are produced by plants, bacteria, and fungi from glucose and ammonia. A human needs 50–75 grams of protein per day. But it is not only the quantity of amino acids that matters. The exact amino acid composition is also very important. We cannot live on just 75 grams of glutamic acid a day. The only way to obtain exactly the right quantities of each amino acid would be cannibalism. The preferred alternative is a varied diet containing enough animal food. Milk and eggs can provide almost the right quantities of every amino acid. We also have a certain degree of flexibility in our amino acid requirements: a surplus of one amino acid can be turned into the amino acids of which there are a shortage. But certain amino acids must be taken up as such from our food. These amino acids are called *essential* amino acids.

For humans, it is easier to obtain a complete package of amino acids from a mixed diet than from an all-plant diet. Deficiencies will lead to degradation of our body's protein, such as muscle tissue. The opposite is widely known in the world of bodybuilding: practitioners may consume food supplements that contain specific amino acids that promote muscle growth. Unlike fat or carbohydrates, amino acids can hardly be stored—our body needs them every day.

The differences in amino acid composition lead to different proteins. There are many types of protein and they all have specific properties (and it is futile to try to address them all within the limits of this section). We will concentrate on what is useful to know with regard to the understanding of flavor. For that reason, we will discuss the role of proteins with respect to mouthfeel and flavor richness.

As far as the mouthfeel of proteins is concerned, the structure of the protein and the ways that this structure can be altered need to be discussed in more detail, especially the unfolding or denaturation and coagulation of proteins. The primary structure of a protein is the result of the (combinations of) amino acids that are connected and the way they are bonded. There are two basic chains of bonds: those that are circular in a coil or helix or linear in a folded shape similar to pleats. Heating and other cooking processes can cause these bonds to unfold: they denature. This is the technical explanation of what happens, for instance, to collagen. Collagen is the most abundant protein in mammals. It is the connective tissue that keeps muscles and other organs together and is present in every small part of the body. In muscles, which are rich in collagen anyway, the forming of collagen is related to the amount of work that muscle needs to do: the more it is used, the more collagen there is. Collagen is a strong helix that will unfold or denature if enough heat is applied. It will transform into gelatin. The effect on mouthfeel is clear: collagen in its pure form is rather hard and elastic, while gelatin is soft and coating. So the first conclusion is that muscles that have done quite a lot of labor need a lot of temperature and enough time to become flavor rich and coating.

Culinary life would be easy if there were no other proteins and processes involved. Proteins are also likely to form new bonds and become solid permanently. This process is called coagulation and is known to everyone who has ever observed the consequences of boiling an egg. Within minutes, the egg white is coagulated—hard and dry forever. With eggs, this process is quite easy to manage. Unfortunately, the proteins of meat within the muscle that are kept together by collagen are likely to coagulate almost as quickly as egg white and do that at the starting temperature that the collagen needs to become soft (60°C). This is the real challenge in the preparation of meat: the outside of much used muscle tissue needs a high temperature to eventually become gelatin, although the inside of the same muscle cannot support high temperatures and turns dry. Fish is easier in that respect. Fish "meat" contains very little collagen because fish float in water and hardly need collagen. The cooking temperature of fish is generally much lower compared to that of meat. The culinary challenge with fish is to maintain its moisture. The proteins are less protected and turn dry more quickly. We will elaborate on this in the chapter on food preparation (Chapter 4).

Not only temperature can denature or coagulate proteins. Acids or salt also have this capacity. This is the essence of cold cooking and the basis of, for instance, carpaccio, ceviche, and dry-cured hams.

Proteins may also have a decisive effect on flavor richness. The already mentioned Maillard reaction (associated with the aromas of roasting meat, almonds, or coffee beans, for example) that is so important in the world of food is a complex biochemical reaction between proteins and carbohydrates at high temperatures. This transformation of molecules (the Maillard reaction) enhances flavor richness. But that is not all. Amino acids have flavor of their own. Incidentally, the first amino acid to be identified was asparagines (in 1806). It was found in asparagus, hence its name. The amino acid that is especially celebrated because of its flavor is (free) glutamic acid, discussed earlier. Umami in itself illustrates the role proteins and their building blocks may have on flavor richness. It is important to note that glutamic acid can only contribute to flavor if the protein contains this amino acid to start with (most, but not all, proteins do but not in the same amounts). Then enzymes are needed to release them.

This section must end with some comments on genetics and genomics—if only to point out that organisms, and more specifically the proteins that they are made of, have their mysterious ways of adapting to nature. These adaptations lead to specific differences in all organisms, plants, and animals. This diversity may be a reason we select particular ingredients for specific purposes. For example, grapes grown in a specific area lead to a distinct wine, milk in another region may lead to a specific cheese, and the properties of the meat from a cow that has grazed in mountains may be quite different than other meats. All through the ages, people have studied positive and profitable characteristics of organisms that we use for our consumption. These particular organisms are then selected and grown. In other words, we have always tried to adapt organisms for our own benefit. And we have gotten better and better at it, certainly now that DNA, genome, and genes are better defined. We are able to locate the genes that are responsible for positive and also for negative characteristics. The latter can be eliminated from the DNA resulting in—for example—plants that need not to be treated with chemicals to prevent diseases. This is what genomics is about and, as far as I am concerned, this is a good step forward.

SUMMARY

In this chapter, the focus has been on the objective side of gastronomy: flavor. For a long time, addressing flavor was hampered by a lack of words. We need objective terms to describe phenomena. This is a prerequisite in communication. The words that we use to describe flavor are mouthfeel and flavor richness.

Mouthfeel is the sensation we experience in the mouth from a product. Essentially, we make a distinction between contracting and coating. Acids and salts cause the mouth to contract. Fats and sugars are likely to leave a layer behind (coating). The absence of contracting or coating substances is called neutral. Neutral can become dry in many ways—for instance, by absorption, preparation, or tannins.

Flavor richness is about the volume or intensity of flavor and about the flavor type. How is the flavor characterized? Fresh and ripe are distinguished. Apple, cucumber, chives, and lemon are examples of "fresh."

Vanilla, mushrooms, garlic, and rosemary are examples of "ripe."

Based on these parameters, the flavor styles cube model can be constructed. Everything that we eat or drink finds its place somewhere in this model. The flavor styles are called neutral, fresh, round, balance low, robust, pungent, full, and balance high. Flavor profiles are dynamic. Differences in (for example) serving temperature, ripeness, or preparation can lead to changes in the flavor profile. Next, the role of the basic tastes in their relation to the flavor styles is discussed. Salt and sour are basically contracting. Sweet is coating, and bitter may lead to dryness. Umami is especially known for its role as a flavor enhancer; it is also coating. Salt and sour are also flavor enhancers, in some respects.

The role of the four big food molecules is also reviewed. They all have specific characteristics and play an important role in food and beverage preparation. It is useful to be aware of their basic qualities. Water or moisture is significant for mouthfeel. One of the challenges of food preparation is to prevent foods from drying because of lost moisture. Water is also a solvent and contributes to flavor richness in that aromas are dissolved in it. Lipids are not soluble in water but are also carriers of aromas. Furthermore, they are an important source of coating mouthfeel. Nevertheless, fats suffer from a bad image, although there is little support for this negativity. In the discussion of carbohydrates, the focus is on starch; the role of sugars has already been mentioned. Starch has several dimensions. In the form of crusts such as puff pastry, it produces a drying effect by absorbing saliva. It can also be used as a binding agent, and it contributes to a coating mouthfeel (such as in sauces). Proteins are the most difficult and challenging of all food molecules. There are many types, all with specific qualities. Collagen may turn into gelatin, although others coagulate and become dry. It is essential to understand food preparation and vinification techniques in order to appreciate how the desired flavor style can be attained.

APPENDIX: HOW LOW-FAT FOODS
GET THEIR TEXTURE[24]

Chemical & Engineering News[25] put together this useful list about thickening and gelling agents in processed food:

Alginates are derived from brown seaweeds. They are used for thickening, stabilizing, gelling, and film forming in foods such as cream and fruit fillings, salad dressings, ice cream, low-fat spreads, restructured meats, and yogurt.

Carrageenans are carbohydrates extracted from red seaweeds. Used for gelling, thickening, and stabilizing, they are often found in ice cream, coffee whiteners, cottage cheese, and low- or no-fat salad dressings. They are also used to suspend cocoa in chocolate milk.

Microcrystalline cellulose comes from tree pulp. It forms a stable gel that provides creaminess and cling to salad dressings, sauces, batters, fillings, icings, and low-fat sour cream. It prevents fried foods from becoming soggy and helps stabilize whipped toppings and chocolate drinks.

Methylcellulose and hydroxypropyl methylcellulose are derived from tree pulp. They provide thermal gelling properties that reduce oil uptake in fried foods and improve the texture of meat alternatives. They can be used to improve the mouthfeel of sugar-free beverages and reduce milk fat in whipped toppings and desserts. A new use is to help trap air in gluten-free foods.

Cellulose gum is made from tree pulp and cotton. It helps retain moisture in frozen dough, tortillas, and cakes and reduces fat uptake in doughnuts. It stabilizes proteins in protein drinks and replaces texture lost when reducing sugar in beverages. Cellulose gum adds viscosity, flow, and glossy appearance to low-fat sauces.

Gelatin is derived from the collagen in pig and cattle skins and bones. It is used as a gelling agent, stabilizer, thickener, and texturizer in desserts, yogurt, and low-fat foods.

Guar gum, a polysaccharide, comes from the seeds of the guar gum bush, *Cyamopsis tetragonolobus*, which is an annual leguminous plant that originated in India. As a thickener, it is eight times more powerful than cornstarch. It controls moisture and adds texture to baked goods. It also controls viscosity in dairy drinks, salad dressings, and condiments.

Pectin is extracted from the peels of citrus fruits and from sugar beets. It is used for gelling, thickening, and stabilizing food. Pectin derived from sugar beets does not form a gel but is used for stabilizing and emulsifying. Pectin is used in jams, jellies, fillings, and confectioneries. It can also be used to thicken and stabilize fruit- and milk-based beverages.

Starch is generally derived from corn, potatoes, or tapioca. Food makers use native and modified versions. Starch can be hydrolyzed into dextrins such as maltodextrin. Starches are used as thickeners, stabilizers, and fat replacers in puddings, sauces, and salad dressings. They are often added to grain-based foods such as breads, cereals, tortillas, and pasta.

Xanthan gum is made by industrial fermentation of sugar by the bacteria *Xanthomonas campestris*. Used in small amounts, it adds viscosity and cling to salad dressings and sauces. It is also used in egg substitutes and in gluten-free baking.

NOTES

1. Heisenberg, W. (1973). *Physics and beyond: encounters and conversations*. New York: Harper & Row, p. 63.
2. Bartoshuk, L. M. (1978). *History of taste research*. In Carterette, E. C. and Friedman, M. P. *Handbook of perception*, volume VIA, Tasting and smelling (pp. 3–18). New York: Academic Press.
3. Kallithraka, S., Bakker, J., Clifford, M. N. Vallis, L. (2001). *Correlations between saliva protein composition and some T-I parameters of astringency*. Food Quality and Preference 12:145–52.

4. Gawel, A., Iland, P. G., Francis, I. L. (2001). *Characterizing the astringency of red wine: a case study.* Food Quality and Preference 12:83–94.

5. Drewnowski, A., Gomez-Carneros, C. (2000). *Bitter taste, phytonutrients, and the consumer: a review.* Am. J. Clin. Nutrition 72:1424–35.

6. Rolls, E. T., Critchley, H. D., Browning, A. S., Hernadi, I., Lenard, L. (1999). *Responses to the sensory properties of fat of neurons in the primate orbifrontal cortex.* Journal of Neuroscience 19(4):1532–40.

7. Guinard, J-X., Mazzucchelli, R. (1996). *The sensory perception of texture and mouthfeel.* Trends in Food Science & Technology 7:213–9.

8. Guinard, J.-X., Zoumas-Morse, Chr., Walchak, C. (1998). *Relation between parotid saliva flow and composition and the perception of gustatory and trigeminal stimuli in foods.* Physiology & Behavior 63(1):109–18.

9. Gilbertson, T. A., Damak, S., Margolskee, R. F. (2000). *The molecular physiology of taste transduction.* Current Opinion in Neurobiology 10(4):519–27.

10. Carstens, E., et al. (2002). *It hurts so good: oral irritation by spices and carbonated drinks and the underlying neural mechanisms.* Food Quality and Preference 13(7–8):431–43.

11. McCleskey, E. W., Gold, M. S. (1999). *Ion channnels of nociception.* Annual Review on Physiology 61:835–56.

12. Schiffman, S. S. (2000). *Intensification of sensory properties of foods for the elderly.* Supplement to the Journal of Nutrition 130:927S–30S.

13. Fuke, S., Ueda, Y. (1996). *Interactions between umami and other flavor characteristics.* Trends in Food Science & Technology 7:407–11.

14 Lustig, R. H. (2010). *Fructose: metabolic, hedonic, and societal parallels with ethanol.* Journal of the American Dietetic Association 110:1307–21.

15. Adler E., Hoon, M. A., Mueller, K. L., Chandrashekar, J., Ryba, N. J. P., Zuker, C. S. (2000). *A novel family of mammalian taste receptors.* Cell 100:693–702.

16. Chandrashekar, J., Mueller, K. L., Hoon, M. A., et al. (2000). *T2Rs function as bitter taste receptors.* Cell 100:703–11.

17. Drewnowski, A., Gomez-Carneros, C. (2000). *Bitter taste, phytonutrients, and the consumer: a review.* American Journal of Clinical Nutrition, Vol. 72, No. 6, 1424–35.

18. Yamaguchi, S., Ninomiya, K. (1998). *What is umami?* Food Reviews International 14:123–39.

19. Schlichtherle-Cerny, H., Grosch, W. (1998). *Evaluation of taste compounds of stewed beef juice.* Z Lebensm Unters Forsch A 207:369–76.

20. Oord, A. H. van den, Wassenaar, P. D. van. (1997). *Umami peptides: assessment of their alleged taste properties.* Z Lebensm Unters Forsch A 205:125–30.

21. Yoshida, Y. (1998). *Umami taste and traditional seasonings.* Food Reviews International 14:213–47.

22. Rolls, E. T., Critchley, H. D., Browning, A. S., Hernadi, I., Lenard, L. (1999). *Responses to the sensory properties of fat of neurons in the primate orbifrontal cortex.* Journal of Neuroscience 19:1532–40.

23. www.hsph.harvard.edu/nutritionsource/what-should-you-eat/fats-full-story/index.html

24. www.npr.org/blogs/thesalt/2011/11/03/141977384/how-low-fat-foods-get-their-texture

25. Bomgardner, M. M. (2011). *Call in the food fixers.* Chemical & Engineering News 89(44):13–19.

CHAPTER 4

Flavor in the Kitchen

NO BEAST IS A COOK

Gastronomy has deep evolutionary roots. No other species in the animal world has developed such a profound interest in his food and done so much about it as humans. Food is life. Therefore, judging quality and quantity of potential food substances is extremely important in the whole animal kingdom. Even the lowest animals—such as corals and sea anemones—have an increased concentration of nerve cells and nerve fibers around their mouth as compared with the rest of their body. Other sedentary animals such as mussels and oysters, which filter their feed from the water, likewise have a high concentration of chemical sensory cells around their mouth, and what little brain they have is located close to the mouth.

As soon as animals start to move about in pursuit of their food, the concentration of senses increases at their front end more than elsewhere: a head develops and can clearly be distinguished from the outside—even in a snail, to mention just one example. The chemical senses keep their function of identifying and assessing the food that is about to enter the mouth but, additionally, the olfaction (smell) helps animals locate potential food sources. This function is supported further by the development of seeing and hearing, all concentrated on the head. The brain—the most important part of the nervous system—is located close to this great concentration of senses: inside the head.

In evolution, the formation of this concentration of senses and nerve tissue is called cephalization—the evolution of a head. Of course, it is also about other forms of interaction with the environment: identifying and avoiding dangers. But it is mainly about food. In his fascinating book, *Catching Fire*, Richard Wrangham[1] is quite convincing in stating that cooking in itself completely transformed the human race. No other animal is a cook. Cooking, or rather preparing food, heating, and transforming it made us what we are. And this process is still ongoing.

Thousands of years of experience slowly but surely taught us how to use nature to our benefit. We domesticated and perfected plants and animals to produce what we like and/or need. We discovered the elaborate process of treating the berries of the coffee plant in such a way that we can brew the widely loved beverage from them. And we know how we can make beer or whiskey out of barley and other grains or how to convert grapes into wine. We have become skilled in working with proteins, carbohydrates, and fats in all kinds of ways. We have learned—the hard way—which, why, and how foods can make us ill or healthy. And throughout the ages, eating and drinking have given us pleasure. Gastronomy has helped to forge friendships and treaties.

The end results of all these developments are fully stocked supermarkets and a huge array of cookbooks and television series. Modern man, in this affluent society, is required to make choices that are hardly needed when food is scarce or only regionally available. Having to make choices is a wonderful feat and, at the same time, a nuisance. We want to make the right choices, and choosing the one means skipping the other. Furthermore, we want to live our lives to the max and not spend too much time in shops or in the kitchen. Cooking today is challenging, to say the least.

PRACTICE OR PURPOSE?

Human evolution has given us a huge array of techniques that we use to prepare our food. Originally, there were several reasons:

- Preservation
- Digestion
- Palatability

Preservation is very functional if you want to eat foods out of season. Before cooling and freezing possibilities were available, other means had to be found to prevent ingredients from perishing. Turning fruits and vegetables into dried fruits, jams, marmalades, pickles, and chutneys, just as drying, curing, cooking, and/or smoking meats and fish, are all examples of early techniques that enabled us to conserve foods. It gave us hams, sausages, stockfish, smoked salmon, ketchup, confits, rillettes, and many other culinary beauties that we can eat long after harvest, catch, or slaughter.

Another important reason for some kind of preparation is to make foods edible or at least easier to digest. Many foods are utterly edible when raw: fruits, tomatoes, cucumber, oysters, and so on. Think also of steak tartare and of sashimi in Japanese cuisine. Bear in mind that the sashimi chef is the highest in Japanese chef hierarchy. Eating sashimi of fugu, the puffer fish that is deadly poisonous if not treated correctly, is a Japanese delicacy that can only be served by certified chefs that have undergone rigorous training. Most foods, however, need preparation. Cassava is an important source of carbohydrates in the tropics and a major staple food of many developing countries. Yet it contains cyanide and needs to be properly cooked to detoxify. In fact, many foods are digested better after they have undergone some kind of preparation. This preparation is needed to help our metabolic system give us enough energy in a relatively short time. Our digestion has problems doing that if we eat only raw foods. This is shown in a study where people ate up to five kilos of raw food for twelve days. It gave them enough calories and it also had health benefits, but all lost weight and had energy problems.[2] We are not like animals: we need our food to be cooked in some way.

However significant the aforementioned reasons are, they are always a function of palatability, which must be considered—above all—as *the* reason for food preparation. Humans like their food to be tasty and have been known to go far in reaching this goal. Palatable foods give pleasure. Pleasure is a personal emotion, which implies that palatability is subjective by definition. Nevertheless, there are foods and drinks that are liked by many people all over the world, and we also know the English proverbs "hunger is the best sauce" and "hunger is good kitchen meat." One of the missions of the science of gastronomy is to learn more about the factors that evoke or enhance pleasure. This understanding can potentially be used in many areas. The most obvious relevance is in the commercial world of food: restaurants and the food industry at large. Less evident is the study of food liking by elderly people in nursing homes, hospitals, and so on. Why do some older or sick people eat too little? Have they just lost appetite, or do we have to work on the palatability of the foods that they are served? Another unexpected angle of gastronomy are the problems of craving, overeating, or the development of a dietary pattern that may be very palatable but not very healthy. Changing behavioral food patterns pose an interesting gastronomic challenge. Chapter 7 of this book is devoted to this subject.

Indeed, the scope of the science of gastronomy is quite wide and goes beyond "the art and practice of cooking and eating good food" (*Oxford Dictionary*). Nevertheless, that is the focus of the present chapter. Factors that influence palatability are discussed in Chapter 7. In this chapter, the spotlight is on food preparation; in the next, it will be on the production of beverages such as wine and beer. The focus is on why we do what we do, what happens in the process, and what are the effects on flavor.

Generally, we do not think about these questions. Most of what we do in the kitchen is a routine activity. We apply all kinds of cooking techniques without asking why. It is a general characteristic of crafts: techniques are handed down from the master craftsman to the apprentice, from generation to generation. In the process, the tricks of the trade are passed down and mostly kept secret. This normally ends when science moves in and explains why and how things happen. It took quite some time before science took a real interest in food preparation. First, the food industry profited from science and applied scientific research. In 1984, Harold McGee published his first version of *On Food and Cooking: The Science and Lore of the Kitchen*. This book was probably the one that opened the eyes of many chefs. Ferran Adrià from Cataluña, Spain (El Bulli restaurant, three-star Michelin closed in 2011), is likely to go down in history as the chef who ultimately brought science to the best kitchens with what has been baptized as "molecular cuisine."

Known or not, all techniques have a purpose. Paying attention to cooking means paying attention to the reasons why we prepare food. That reason is none other than to improve our foods, to make them more easily digestible or more attractive because they are not or are less so when they are raw. Therefore, we will look at food preparation itself. How does flavor come about, what are the characteristics of the most important techniques that we use to prepare foods? What do we need to consider when combining different flavors? Perhaps this approach gives another perspective. Let us start at the beginning: the choice of the ingredients. Their characteristics ultimately determine what the chef can or should do.

INGREDIENTS: BASIC QUALITIES

Agriculture is a human invention. It is not natural, yet it is the basis of modern civilization. In its quest for food, the human race has learned a lot about growing the ingredients we need and desire. The achievements in this field are quite impressive and have enabled the growth of world population. We would have starved otherwise. Selective breeding is a part of the secret; finding the right spots and methods to produce our nutrients is another piece of the puzzle.

Agricultural land is scarce. Table 4.1 shows that one-third of the land area of the world is classified as agricultural. The rest consists of cities, mountains, deserts, swamps, permafrost, and so on. Less than one-third of the agricultural areas (or about 10% of the total land area) can be cultivated and produce useful plant products. The remaining two-thirds of the world's agricultural ground is suitable for ruminant animals because it is covered by grass, shrubs, or other plants that only they can digest. There is a place for animals and plants in agriculture and, in light of the growing world population, there is every reason to be prudent with the available agricultural land. Put culture back in agriculture, and think twice before using precious land for other purposes such as the production of biofuels.

TABLE 4.1
Agricultural Land in the World

Characteristics of Agricultural Land in Various Geographical Regions				
Geographical Region	Total Land (1,000 sq. miles)	Land That Is Agricultural (%)	Part of Agricultural Land Cultivated (%)	Part of Agricultural Land Permanent Pastures (%)
Total world	**50,495**	**35**	**31**	**67**
Developed countries	*21,176*	*36*	*33*	*66*
Developing countries	*29,319*	*34*	*29*	*69*
Africa	8,994	37	19	79
Asia	10,334	38	45	53
Europe	1,826	49	55	38
Oceania	3,254	61	9	91
North America	7,084	27	46	53
South America	6,771	31	15	81
USA	3,524	47	43	56

Source: FAO Production Yearbook, www.ansi.okstate.edu/breeds.

There is an intriguing relation between the land and the circumstances, the people who use it and the products they produce. The legislation on the Appellations of Origin is based upon the recognition of specific and regional circumstances that lead to differences in natural products. Appellations are most known in the world of wines but may apply to many other ingredients and products as well, such as cheeses, fruits, and other regional specialties. The general rationale is that the local circumstances and techniques are so special that the products that are grown and made in these specific and well-defined areas can be recognized. Local and/or national governments take an interest in defending these products. These days this interest has developed an extra dimension because of the negative effects of globalization and food miles. More and more people seem to take an interest in local foods and their producers.

Every ingredient has specific qualities. These form the basis for the chef. For the best results, he needs the best ingredients. It is a one-way street: poor ingredients or ones that are not perfect for the desired result will not lead to success. However, even the use of perfect ingredients will not guarantee that the final result will be good. Much can go wrong in the process of preparation. Most recipes feature a list of ingredients that are needed. Rarely are the desired characteristics of these ingredients specified. Nevertheless, this can be the reason for success or failure in the preparation. Take an apple, for instance. There are many different varieties, and most of them have distinct properties. Consequently, some are better for preparing applesauce and others are better in a fruit salad. Therefore, the best gastronomic books, including the ones on wine, will point out what specific peculiarities make the difference.

Ingredients may look the same but might be very different. Circumstances influence the presence and quality of the most important food molecules, such as water (juice), lipids, proteins, and carbohydrates. And not only those—there may also be variations in texture, acidity, vitamins, minerals, and other factors that determine nutritional value. Consequently, differences in the ingredients will lead to different and sometimes unexpected outcomes after preparation. Knowing about the composition of ingredients is a part of understanding the mysteries of successful preparation. The following influences on the development of flavor will be discussed in more detail:

- Climate and location
- Cultivars, varietals, and breeds
- Production methods
- Ripeness

THE INFLUENCE OF CLIMATE AND LOCATION ON FLAVOR

In discussing flavor, the first things that likely come to mind are the influence of the climate and the quality of the soil. Every climate has its own peculiarities, limitations, and opportunities, leading to specific products. Hawaii has its pineapples, and Burma has rice. Brazil, Indonesia, and Ethiopia are known for their coffee, China and India for their tea, while Ivory Coast and Ghana are producers of cacao. Italians are known for their tomatoes and wine. The French can also be proud of their wines, just as of their cheeses. Wine grapes are even cultivated in the Netherlands, but vegetables in general—in all varieties and in huge quantities—are what the Dutch can be proud of. We could go on and on mentioning countries and regions with their specialties. These have primarily developed as a result of climatic conditions in combination with the location, which includes the landscape, the soil, and everything else that defines a certain region.

Climate and location are basically two different concepts; both however determine quality and characteristics of the product and ultimately even its presence. Products may thrive in a certain environment, although the weather circumstances are rarely the same. One year there is too much rain, the next year too much sun, a year later there will be a late frost, or the plant might be harmed by hail. No year is the same; however, some regions are more stable than others.

On the other hand, the landscape and composition of the soil are more or less a given fact. There is a certain altitude; the land may be close to the sea or behind a mountain ridge. Within the same climatic region this can, for instance, influence the amount of rain, the amount of light (luminosity), or the difference in day and night temperatures that influence the photosynthesis and the development and retention of aromas in fruits.

And then there is the type of soil. Soil is basically made up of finely ground or eroded rock particles and a top layer of organic material or humus. The composition of the soil can be classified according to the size of particles such as sand, silt, or clay. Of these, sand is the largest and determines aeration and drainage characteristics. The more sand, the less the soil is capable of retaining water. The tiniest particles, clay, can bind water. Furthermore, they contain minerals, which are plant nutrients. Pure clay is not permeable. Soil may be mixed with larger pieces of rock such as gravel, pebbles, chalk, quartz, granite, or slate, each with their own characteristics that determine if a certain plant will grow well or not.

Soil can be more or less fertile, but that does not necessarily mean that rich grounds are always to be preferred over less fertile ones. In fact, if vineyards are planted on rich soils, the plants will grow vigorously and mostly produce new sprouts and leaves rather than fruit. It is on the less fertile soils that the best grapes are grown. The availability of water also has a big influence, and again not too much, and certainly not too little. Water stress obstructs the photosynthesis and, therefore, the production of sugar. Water is also needed to transport nutrients from the soil to the plant. Just to mention one example: potassium moves water into the cell of the developing fruit. Thus, if you want juicy tomatoes, for instance, adequate potassium must be available.

Conditions are one consideration; how people make use of them is quite another matter. The soil needs to be worked and can be treated with natural or industrial fertilizers. Unfortunately, this is not always done properly. The best farmers, however, will always stress the importance of a well-worked and vital soil. They will try to get the most out of their location and their circumstances.

THE INFLUENCE OF CULTIVARS, VARIETALS, AND BREEDS ON FLAVOR

Within families of products, there are varietals that have a specific character. In most cases, these are the result of breeding desired characteristics. Flavor can be one of them but not exclusively. Conditions that involve growth can also be the reason. Some varieties of a plant thrive in colder areas, others in warmer; some are not as susceptible to frost, wetness, or drought; some need acidic soils, others need calcareous soils, or clay, slate, or sand—just to mention some examples. In the world of wine, all these varietals are quite well known and so are their specific characters. New agricultural areas look at places in the world with similar circumstances, which helps determine what varietals to plant.

Evidently, it is not only in the wine world that varieties get a lot of attention. In all living organisms, there are these differences. The basic characteristics of the product determine how the product can be used in the kitchen. One type of potato is different from another type of potato. There are differences in flavor and texture. Therefore, one type of potato might be used for boiling, another type for French fries, and yet another type for baking or making crisps. In other products, differences are crucially important as well. If you bite into four different types of apples or tomatoes, you will notice differences in acidity and texture. And think about fish and meat. Salmon, cod, red snapper, and Dover sole all have their peculiarities and culinary possibilities. In meat, some breeds of cattle are famous for the type of beef they produce, such as Limousin, Charolais, Blonde d'Aquitaine, and Black Angus. Other cattle, such as the Friesian Holstein cows, may be world champion milk producers, but their meat is nowhere near as flavorful. And as far as milk is concerned, the milk of cows is markedly different compared to goat's or sheep's milk.

Scientifically, the development of new varieties is an advanced field. The ideal is to produce plants and animals that are higher-yielding, resistant to pests and diseases, drought-resistant, or regionally adapted to different environments and growing conditions. The positive side of this is clear: it brings stability and enables mass production. The downside is that interesting but low-yielding "difficult" varietals tend to disappear. Biodiversity declines, much to the dismay of the gastronomic world. Ideally, farmers and chefs have good relationships—or rather keep on having them. This keeps production and demand well in hand and distribution lines as short as possible. Furthermore, the dialogue between chefs and producers stimulates mutual understanding for each other's dispositions.

The Influence of Production Methods on Flavor

The days are long gone when the world population could be fed on natural resources. There are still animals (game), plants, grains, nuts, fruits, and herbs that can be harvested from nature. And there remain even a few items that are only found in the wild and to which no cultivated alternatives exist. There are a few types of mushrooms, for example, that cannot be successfully grown yet, but these are the exceptions. Basically everything that we eat is produced for that purpose and more or less intensively. Plants and animals are supplied with nutrients that somebody thinks are best for one reason or another. This is the big difference with eating from nature. In nature, living organisms have to source their own food. If they are successful in doing so, they are healthy and strong. If they have problems or fail to find the nutrients that they need, they will perish and die; ultimately they become extinct.

Evidently, this is not what we want. We would like, and do effectively need, quite a lot of food for ourselves, and we expect nature to deliver our demands—in quantity and in quality. In fact, we want something rather unnatural from nature: we want stability, and we go quite far in accomplishing that. So we have selected the best cultivars and breeds and have learned what is needed to get desired results. The famous chickens and pigeons of Bresse are raised with a carefully planned protocol that includes living space and a specific diet, predominantly containing grains.

Food, circumstances, and way of life have a strong relation with flavor. Grains, in general, have a positive effect on the quality of meat. The combination of ample food and little exercise tends to promote fatness. In the world of meat, this combination implies tenderness, juiciness, and flavor. The muscles of animals that are required to do a lot of work develop connective tissue that is rich in a protein called collagen. It is the connective tissue that keeps muscle fibers together. We have seen in Chapter 3 that collagen needs time and temperature to change into gelatin. This is a culinary challenge that can easily go wrong and lead to dry meat that falls apart. Meat with little collagen is tender and easier to prepare. This gives the farmer a strong interest in limiting the activities of the animals. These animals will also burn less energy, so they require less food.

In the world of meat animal farming, many production methods were developed to obtain meat that was cheap and tender. Those two qualities are not easily matched, and these days the systems that are needed to achieve these goals are frowned upon. And indeed, the images of mass meat production are not pleasant to see and are far from natural conditions. However, the applied methods fit the consumer demand for inexpensive and tender meat. If we want better circumstances for the animals and meat that is more flavorful, we need to pay the price. A change in consumer behavior could also help. To paraphrase Michael Pollan: Eat food. Not too much. Mostly plants.[3]

Supplying the necessary nutrients is not the only role that we humans have taken over from nature; we also have grown very efficient in protecting plants and animals from all kinds of diseases. Chemical herbicides, pesticides, and fungicides, just like hormones and antibiotics, can make life easier for farmers. Others choose not to use them and work organically. They choose to use nature to their benefit. They may plant certain flowers or herbs that are valuable to the soil or ones that attract desired insects that serve as predators. Or they may attract certain birds that love the insects the farmer does not like. Or farmers may spread pheromones to sexually confuse damaging insects, thereby reducing their numbers.

Organic agriculture is gaining popularity. To work organically or not is a fundamental choice that leads to debate, and it ultimately becomes a marketing issue. In many cases, flavor is used as an argument, but in practice, the organic alternatives do not always perform as expected. Critics claim that yields are too low and, therefore, organic agriculture could never be a viable alternative to modern high-tech agriculture—especially in view of the growing world population. This remains to be seen. New studies show that a better selection of varieties could much improve the performance of organic production.[4,5]

All the same, organically produced food is not an implicit guarantee of quality. Extrinsic factors may play a role: organic fruits and vegetables do not always look as nice as their less natural counterparts, which may very likely have a negative impact on flavor perception. More fundamentally, the question may be raised as to whether flavor is the most important argument in discussing the issue. Preserving the planet and life are essential. How are we going find the balance between ample food production for billions of people and the use of resources and means? Leaving these big questions aside and returning to the core of gastronomy: however something is grown, it is always important to look at the desired product qualities and make choices accordingly.

THE INFLUENCE OF RIPENESS ON FLAVOR

It is gastronomy which decides the point of fitness of every nourishing substance; for not all of them are desirable in the same way. Some of them must be used before having reached their final stage of development, like capers, asparagus, suckling pigs, squabs and other creatures which are eaten in their early days; others must be enjoyed at the moment when they have attained the peak of their growth, like melons, most fruits, mutton, beef, and all adult animals; still others when they begin to decompose, like medlars, woodcocks, and above all pheasants; others finally, when all the arts of gastronomy have removed their noxious qualities, as with potatoes, the cassava plant, and others.[6]

Ripeness has a big impact on flavor, and this quote from Brillat-Savarin points out interesting differences. The relation between flavor and ripeness is probably best known in fruits. A pear, melon, plum, peach, or mango is best consumed at maximum maturity. It is then that they are rich in flavor and succulent, coating in mouthfeel. We would rather not eat them when they are still hard. Note, however, that a ripe green apple will always be hard and contracting even at its peak. If it turns soft, it has lost its appeal and will probably be thrown away. Ripeness has several dimensions, and every category has its own ideal ripeness. It is an art to choose the right moment for consuming the product. There may be a reason to select a moment before maximum ripeness. For instance, if we want to use strawberries in a fruit salad or as a garnish, we want to be able to slice them. A strawberry that is not totally ripe may in that case be a better choice. When making strawberry jam, we would rather have the strawberry ripe or even overly ripe.

Modern society has not only forced its will upon the production of our food, it has also influenced decisions on the ripeness. Mass distribution requires foods to have sufficient "shelf life." The solution is to harvest early, not at full ripeness. This is not without consequences. After maturation, a fruit or vegetable keeps on developing flavor, especially sweetness and aroma compounds. When it is harvested early, this development partly stops. When fruits or vegetables have kernels, this can easily be spotted: the pips are still white. At full maturity, these would be dark brown. In flavor, they will be less sweet, more acidic, and not as rich as the ripe fruit would be.

The so-called nonclimacteric fruits stop ripening directly after harvest. Examples are grapes, cherries, cucumbers, lemons, and pineapples. Climacteric fruits continue to ripen after harvest. These fruits can be ripened naturally or artificially. Examples are bananas, apples, melons, avocados, and tomatoes. Traders are well aware what the ideal conditions are to store these kinds of fruits and vegetables. Temperature and humidity have a big impact. At home, some vegetables and fruits are better not kept in the fridge, while others need to be kept cool. Storage and ripening influence the flavor.

With meat, it is a different story altogether. Immediately after the slaughter rigor mortis sets in. A complicated chemical process causes the muscles to stiffen up. This phase takes (depending on the kind of animal and the temperature of the surrounding area) about five to 24 hours. During this period, all meat, including fish, is not very attractive. It is hard and tough. Then the aging process sets in; enzymes start working on the proteins, which makes them much better to consume. In fish and meat of young animals, this process does not take long. Depending on the animal and its age, within a little more than a day and up to a week flavor richness and texture has reached its desired level. The meat of mature animals such as beef, which is kept at temperatures of between 1°C and 3°C, improves substantially during a period of about three to four weeks. The meat becomes noticeably more tender and acquires more flavor.

A disadvantage to ripening is that, for several reasons, the product becomes more costly. These include length of storage and the loss of moisture (especially in meat and cheese), which means that there is less weight to sell. With the aging of meat, discoloration of the meat might occur, so that parts of the meat have to be cut away. Products that might spoil need to be stored in expensive cooling facilities. In other words, it is clear that products that have ripened are more costly than their young counterparts. The investment of time and money is rewarded with flavor. The same is true for cheese, wine, and spirits. They also need to ripen in order to attain their maximum level of quality but get better in the process, which makes the cost of maturing worthwhile.

Ripening (maturing) does not always mean that the quality improves. There are products that lose their quality so quickly after being harvested that the time of consumption needs to be as soon after harvest as possible. Examples are lettuce leaves and types of cress, oysters, mussels, and shrimp. The success of a fish restaurant by the ocean often can be ascribed to the immediate availability of the fish, which almost immediately after leaving the ocean reaches the plate. The flavor of this fish—if prepared by a good chef—will be at its best. And, likewise, there is a tendency among serious restaurants to have their own vegetable and herb garden or a partnership with a farmer that grows for the restaurant.

THE ESSENCE OF COOKING

Every country has good as well as *very good* ingredients. Even then, you can ruin them completely by lacking cooking skills. A great chef will, therefore, always be critical in the selection of his ingredients and their suppliers. Bad ingredients will never lead to something palatable. On the other hand, a chef's skill and taste will enable him to transform even average ingredients into a delicious dish. It is essential to know about and to master techniques. After having elaborated on ingredients, we will now focus on preparation.

Preparing food is all about transforming molecules. That is not very romantic, and some even consider it as unattractive and "not done" to consider cooking from a physical and chemical point of view. Yet that is what it is all about, and all gastronomes in history realized that. Brillat-Savarin himself reproached his chef for serving a sole "pale, flabby, and bleached" and noted how his guests at the table were annoyed about that.

> This misfortune happened because you have neglected the theory, whose importance of which you do not recognize. You are somewhat opinionated, and I have had a little trouble in making you understand that the phenomena which occur in your laboratory are nothing more than the execution of the eternal laws of nature, and that certain things which you do inattentively, and only because you have seen others do them, are nonetheless based on the highest and most abstruse scientific principles. Listen to me with attention, then, and learn, so that you will have no more reason to blush for your creations.[7]

From the perspective of gastronomy today, we will discuss how mouthfeel, flavor intensity, and flavor type can be controlled. Understanding the essence of cooking starts with a basic understanding of the two most significant variables in (normal, traditional) food preparation: time and temperature. The use of a certain amount of moisture can also be an important aspect. Boiling an egg illustrates this. We need boiling water as an agent. At sea level, it boils at 100°C. Now look at time: a raw egg is slimy and not exactly appetizing. After about three minutes, the white of a standard size egg has started to coagulate; it would still be fluid toward the egg yolk. One minute later, the coagulation of the yolk sets in. The yolk will be solid after five minutes but not completely dry. A hard-boiled egg of six minutes is completely dry. If we continue to boil the egg, the yolk will attain a greenish color and become unappetizing. We all know this. In hotels, nobody acts surprised when a guest requests that an egg should be cooked for exactly half a minute. Within three minutes, its mouthfeel changes from (extremely) coating to dry. And this has everything to do with the rise of the internal temperature and the reaction of the proteins to this temperature. Note that these changes are irreversible: once hard, an egg stays hard.

Boiling is not the only way to change the flavor profile of an egg. Eggs can also be fried, scrambled, or poached or used to make an omelet, tortilla, meringue, cake, soufflé, or sauces such as zabaglione, hollandaise, or mayonnaise. This is just for starters. The history of the art of cooking contains thousands of interesting applications. They all have one thing in common: the success of the outcome depends on the control of time and temperature. We need to reflect on the huge changes that have taken place with the egg in a very short period of time. The dynamics are astounding. Once the egg has changed, it cannot change back. Now realize that proteins appear in different guises and not just in eggs. Similar processes take place in the preparation of meat and fish. The skillful cook is a master at controlling temperature and time and understands what he is doing and why.

The molecular transformations take place by using energy in some way or form. In fact, energy and heat are the two universal ingredients of cooking. Energy can elegantly be considered as "the ability to make things happen." Heat is just one of the forms of energy, albeit the most common in cooking. It involves energy that flows from a substance with a higher temperature to another with a lower temperature.[8] This is called heat transfer. There are a few basic methods to make this happen:

- Conduction: heat transfer by direct, physical contact. You can hold your hand above a hot pan for a while, but you get burnt instantly by putting it on the hot surface. The density of the metal is an important factor. A light metal such as aluminum warms up quickly and loses its warmth just as quickly; a cast iron casserole or a copper frying pan takes (much) longer to heat up but keeps temperature much better when food is put in it.
- Convection: heat transfer by the environment, such as steam or air. Movement and type influence the transfer. The heat transfer in a conventional oven goes much faster when forced with a ventilator. The transfer is even more effective with steam. Again, you can easily keep your hand in a hot oven for a while but not over boiling water.
- Radiation: heat transfer by heat rays. These are part of the spectrum that is invisible: microwave and infrared rays are examples. Grilling on charcoal is another illustration. What happens is that the glowing coals emit visible and infrared radiation.

Atoms in the food absorb some of these rays and convert them into heat. The absorption works at its best in dark substances, which implies that as food turns brown, the process speeds up considerably.

In general, lower temperatures (below 100°C) give longer cooking times and are suitable for softening hard textures. Products that are already soft can be prepared at a higher temperature and shorter cooking time. Preparing at high temperature may have an additional effect. Flavor richness will certainly be enhanced when the so-called Maillard reaction takes place. Preparation at high temperature may also give a nice crust, which clearly influences the flavor profile. A good crust also influences the quality perception, as it is directly associated with freshness.

Timing is crucial for a good result. Vegetables or pasta that have been cooked too long lose their texture and attractiveness. Neither are greatly blackened fish, steak, or potatoes very appealing. Most ingredients have an ideal cooking time, and it often is quite precise. One of the arts of cooking is to have everything ready right on time. This demands thinking and planning ahead, especially when more people are involved. They all want to eat at the same time, have the food perfectly prepared, and served at the right temperature.

CHARACTERISTICS OF THE
PRINCIPAL TECHNIQUES

COOKING AND BLANCHING

Cooking and blanching are two sides of the same coin; both involve boiling water. The difference between the two is time. In cooking, the product is kept in the boiling water until it is done. Blanching implies that the product is cooked just for a short time and then plunged into iced water or kept under cold running water with the intention to halt the cooking process. This technique is meant for preparing the product and finishing it later, for example, by sweating it in butter. For making

French fries, the potatoes are first blanched and later deep-fried. Both techniques are mostly used for preparing vegetables, fruits, potatoes, and pasta.

Blanching is also used as a preparation for freezing vegetables. The period of heating is short enough to keep the texture of the vegetables, but most enzymes that might cause undesirable changes even at low temperatures are inactivated.

Special meat products such as kidneys and sweetbreads need to be blanched before they are fried or roasted.

Cooking/blanching (in water at 100°C)	
Effect on Flavor	
Mouthfeel	*Flavor Richness*
Textures soften. The ideal, in most cases, is to keep a certain "bite."	Ideal for keeping the primary flavor of the ingredient.

POACHING

Poaching is a technique that is used to prepare fish, meat, or eggs. Poaching may seem similar to boiling, but there are two big differences:

- In poaching, the temperature stays well below 100°C; often 90°C is used; poaching can also be done at lower temperatures, but not under 60°C.
- Poaching is done in a good stock or sauce with enough concentration to maintain the flavor of the main ingredient. If the concentration of the liquid is too low, flavor components of the fish or meat will go into the liquid. This is called diffusion and is the name of the game when you are making a bouillon, stock, or sauce. The mission of poaching is not improving a liquid but preparing your fish or meat.

Poaching (in stock at max. 90°C)	
Effect on Flavor	
Mouthfeel	*Flavor Richness*
Textures soften. The ideal, in most cases, is to keep a certain "bite."	Ideal for keeping the primary flavor of the ingredient. Enrichment and character has to come from sauces, salts, acidity, herbs, and spices.

STEAMING

Steaming is cooking over a boiling liquid in a closed container. The substance to be cooked lies over the boiling liquid on an underside that allows the hot vapor to go through. The difference with these techniques is that the substance is only in contact with the vapor, which is by definition over 100°C. This high temperature implies that the cooking process can go rather quickly. It is important to pay careful

Steaming (over a boiling liquid)	
Effect on flavor:	
Mouthfeel	*Flavor Richness*
Textures soften. The ideal, in most cases, is to keep a certain "bite."	Ideal for keeping the primary flavor of the ingredient. Enrichment and character has to come from sauces, salts, acidity, herbs, and spices.

attention, especially when preparing fish. There are also pressure steamers in which the process is even quicker.

Steaming is popular in some dietary regimes because no fats are involved. When aromas are added to the liquid, these may give some character to the steamed substance.

FRYING/PAN FRYING/STIR FRYING (BAO TECHNIQUE)

There are different names for a similar principle: cooking at a high temperature with a limited amount of oil or fat. The fat serves as an agent that prevents the products from sticking to the surface. Depending on the type of oil used, it may also serve to increase the temperature. The first function of the fat is to prevent the food from sticking to the pan but not to add flavor. The big exception is frying in butter, which will add flavor. However, the burning point of butter is lower than oil, which implies that butter cannot always be used; especially not in stir frying.

Most frying is done in a pan with a flat bottom; for stir frying, a wok is used—it has a round bottom. For real stir frying—as it is done in Asian cuisines—a great deal of heating power is required. The starting temperature must be at around 140°C. This basically implies that the pan must be preheated well, even before the ingredients are added. Furthermore, this temperature needs to be maintained; therefore, the size of the pan should be in line with the amount that is fried and not too much food should be put in the pan at the same time. If too much is added, the temperature will drop and juices will flow out of the product and ultimately evaporate. You will end up cooking instead of frying. Good frying pans (the more expensive ones) have thick bottoms that can retain quite a lot of heat.

These techniques are meant to be short, intense, and suitable for products that do not need a long time to cook, such as filets of fish or poultry, steaks, and vegetables such as zucchini. Essentially, the aim is to get some kind of browning on the outside, a caramelizing of natural sugars, and (ideally) a crust—the Maillard reaction. Stir frying is one of the main techniques of Chinese cuisine. To speed up the cooking process, the ingredients are cut into smaller pieces. There are two variants:

the Bao and the Chao techniques. In the latter, liquid is added, which is explained here.

Frying/pan frying/stir frying (Bao technique) (in a hot pan with a little oil or fat)	
Effect on Flavor	
Mouthfeel	*Flavor Richness*
The forming of a crust on the outside will give some dryness.	Flavor richness and ripe flavor tones increase because of the Maillard reaction.

SAUTÉING/STIR FRYING (CHAO TECHNIQUE)

These techniques start like the previous ones, at a high temperature and with a little oil of fat, but they end differently. There are a few versions. In the first, a lid is put on the pan, which causes steam to form. In the other variant, a liquid is added. This can be a stock or sauce or also a beverage such as wine or beer. This is done to add flavor, create a sauce, or to prolong cooking times, if needed. The result is that the crust of frying will be lost. In Chinese cuisine, this technique is often used in stir frying and is called Chao.

Sautéing/stir frying (Chao technique)	
Effect on Flavor	
Mouthfeel	*Flavor Richness*
Textures soften. The ideal, in most cases, is to keep a certain "bite"; depending on the sauce, it may become coating.	Flavor richness and ripe flavor tones increase because of the Maillard reaction.

These techniques are appropriate for smaller pieces of meat and veg-etables—whereas "dry" frying can be used to prepare larger pieces.

ROASTING/BAKING

Roasting is the classic technique used to prepare larger pieces of meat. Nuts, seeds, coffee, and cocoa beans also owe their flavor to roasting. The oven is the principal instrument for roasting; however, a roasting pan or rotating spit can also be used. The essence of roasting is dry heat; temperatures are well above 100°C, depending on the tenderness of the meat to be roasted. The less collagen there is, the higher the temperature and the shorter the process can be.

Traditionally, meat was first seared and then put in the oven until it was done. In the first part of the process, the meat is browned and obtains its desired flavor. For a long time, people believed that the searing also worked as a kind of seal that contained the juices. This is not true. Consequently, it is also possible to reverse the order in roasting. In modern cuisine, the meat is first brought to the desired internal temperature and browned later. This way juices are better retained; the roasting process then can be controlled more accurately.

Baking is basically the same as roasting, but the word *baking* is used with regard to bread and pastry. It is generally presumed that, in roasting, temperatures are higher and cooking times shorter than in baking.

Roasting/baking (dry)	
Effect on Flavor	
Mouthfeel	*Flavor Richness*
A crust is formed, which gives some dryness.	Flavor richness and ripe flavor tones increase because of the Maillard reaction.

GRILLING/BARBECUING

Grilling may in some ways also be considered a form of roasting and baking, although the transfer of heat is different. The transfer of heat in grilling is mainly by radiation. Grilling is the most original of the cooking techniques. It can give more flavor richness than the other techniques. This is particularly the case if the grilling is done over charcoal because it produces much radiation but also smoke, which adds flavor richness and some kind of convection. In general, the word "grill" refers to food preparation over or under a source of heat. Temperature is high, and mostly smaller pieces are used for grilling. Grilling cannot really be done in a grill pan. This is actually more like frying because the heat is transferred by conduction. The only thing a grill pan does is give the characteristic gridiron look. The extra flavor richness given by the smoke cannot be expected. Because the heat is high, products that need some time to cook may easily get burnt. This should be avoided because it does not taste good, which is the way nature warns us of the carcinogenic chemicals that are being formed. The solution is to do some grilling for flavor and appearance and then continue the cooking in an oven, or the other way around: grilling precooked products.

It is interesting to note that the words *grilling* and *barbecuing* are often used as synonyms. Originally, they were quite different. The traditional barbecue refers to a slow process with a lid over the gridiron and the charcoal. It is suitable for larger cuts of meat. The meat is partly roasted and partly smoked, which will therefore also contribute to flavor richness.

Grilling/barbecuing	
Effect on Flavor	
Mouthfeel	*Flavor Richness*
Except for reaching a state of doneness, no specific effects on mouthfeel can be expected. There is a risk of dryness.	High flavor richness and ripe flavor tones, especially if wood or charcoal is used.

SIMMERING/BRAISING/STEWING

Simmering, braising, or stewing are again different words for basically the same principle: cooking is done in a closed pot and in a sauce. The goal is to break down the collagen of tough pieces of meat, which indeed happens given enough time and the right temperature. It may take a couple of hours before the process is completed. Ideally, if done properly, the result is a coating dish that is rich in flavor. The collagen dissolves into gelatin and forms natural bonds with the other ingredients. In fact, many of the old-time kitchen classics such as *boeuf bourguignon*, osso buco, and coq au vin are all examples of stewed dishes.

These techniques are particularly suited for less expensive cuts of meat, which means that the added value can be high. Unfortunately, these techniques suffer because many people today spend as little time in the kitchen as possible. Which makes you wonder: the process indeed needs time, but it does not require someone be present. If planned ahead, these kinds of preparations do not take up too much time. Furthermore, the dishes can easily be stored, even frozen, and reheated.

Just as with poaching, the quality and the concentration of the stock is crucial for a good result. If the stock is not well reduced, flavor compounds from the food will move to the stock (diffusion) and a part will also evaporate.

Simmering/braising/stewing (with liquid at low temperature)	
Effect on Flavor	
Mouthfeel	*Flavor Richness*
Soft and coating thanks to the softening of textures and the gelatin that comes free.	Rich flavor, often with ripe flavor tones.

DEEP-FRYING/SHALLOW FRYING

Deep-frying refers to the preparation of food submerged in hot oil (between 160°C and 190°C). In shallow frying, the food has oil up to the sides but is not fully immersed. These techniques cook food extremely fast. The oil is meant to serve as a means of conducting the heat, not to make the food fatty. In fact, if done properly, the fat cannot even enter the food. The moisture within the food will quickly begin heating up. Steam will start forming, keeping the oil out. The food will only get greasy if the oil cools down or if the food is kept in the oil for too long. Products need to be dry before being deep-fried.

The most important result of deep-frying should be the crust. A crispy outside and a soft inside is a desirable characteristic of many foods. As soon as the crust is lost, the food has also lost much of its appeal. Deep-frying is a popular technique in Chinese cuisine, where it is often used in combination with other cooking techniques or with a batter, such as for an egg roll.

Overheating and overusing (deep-)frying oils is a main concern. Rancid-tasting oil and even toxic compounds can be formed. The oil needs to be replaced frequently. This is a disadvantage of the technique: the oil needs to be disposed of. Used oil can be processed into biofuel. Another concern is the flammability of the oil. To extinguish burning oil, water must not be used. A fire blanket or special powders will do the job.

Deep-frying/shallow frying (in hot oil > 160°C)	
Effect on Flavor	
Mouthfeel	Flavor Richness
Crisp outside, soft inside.	High flavor richness, mostly with ripe flavor tones.

OTHER TECHNIQUES TO GET
OR TO KEEP FLAVOR

SEARING, PAPILLOTTE, CLAY POT, *CROÛTE, SOUS-VIDE*

It has been stated earlier in this book: one of the main challenges of cooking is preserving moisture. There are two principal reasons for putting effort into that. The first is that moisture contributes to a pleasant mouthfeel. Juiciness is in most cases preferred over dryness. The second reason is that certain flavor compounds dissolve in water; when moisture is lost, parts of flavor are lost as well. Therefore, it is logical that culinary history has passed down some interesting practices that are essentially meant to retain liquids and their flavors.

First, a long-lasting misunderstanding needs to be cleared up: searing meat is not sealing the meat. Around 1850, the famous German chemist, Justus von Liebig, proposed that meat juices would be locked in by searing the meat. The great chef August Escoffier promoted this suggestion in his work, and so many writers and professional chefs think that it is true—even these days. Yet surely Escoffier must have seen that the crust of a steak is not waterproof. By the 1930s, rather simple experiments showed that Von Liebig's idea was wrong. Just as anyone would suspect, moisture loss is related to the temperature. The higher it gets, the more juices are lost. And even if juices have been retained carefully in the preparation, there is still a danger that lurks. After roasting, it is advisable to allow the meat to rest; count at least half an hour for large roasts. During this time, the meat structure becomes firmer and its water holding capacity increases. When it is carved too quickly, meat juices are likely to flow out of the meat easily, with all the negative effects described earlier.[9]

Some beautiful culinary solutions have been cooked up to solve the juice problem. They always involve preparing food in or with some kind of material that prevents moisture and flavor from disappearing. The first method to mention is what the French call *en papillotte*. The translation would be "in parchment." The Italians refer to it as *al cartoccio*. Fish or poultry, together with some vegetables, herbs, and seasoning, are wrapped in, for example, greaseproof paper, aluminum foil, or large leaves (such as banana, tobacco, lotus flower, grape, etc.). In fact, many materials can be used. Escoffier even used a pig's bladder. In some cases, the leaf wrappings are also meant to be eaten. Whatever is used, the wrap should be folded carefully and have no holes. It is cooked in the oven or grilled on an open fire, so the inside is actually steamed.

The ancient Roman version of this technique is called the Römertopf; a clay pot that, incidentally, still can be purchased. The generic name is "clay pot cooking," and it is found in all kinds of ancient cuisines. An unglazed clay pot is soaked in water prior to putting it in the oven. The steam that is formed stays on the inside and helps to contain juices and aromas.

The same goal can also be achieved by using some kind of dough (i.e., based on salt or puff pastry) in which the food is prepared. When put in the oven, the dough seals and keeps the flavors inside. Classical dishes such as *paté en croûte*, meat pies in general, or beef Wellington are examples. Even the folded pizza *calzone* is an illustration. The *soupe aux truffes noires* of the legendary chef Paul Bocuse is a chicken broth with black truffle and foie gras that is covered with puff pastry and baked in the oven. The pastry cover contains all the wonderful aromas within the bowl. They emerge when the pastry is opened in front of the guest at the table.

Very practical, but more prosaic, is the use of a plastic bag in which food is sealed in a vacuum (or, in French, *sous-vide*). This started out as a way to preserve food but evolved into a cooking technique that is often used in modern cuisine. Besides the food to be cooked, aromatics can be included. The flavor intensity of such dishes is unlikely to be very high but rather subtle and elegant. This technique is also often used as a part of the preparation. Meat or fish is first brought to the desired temperature and then stored. It is reheated (i.e., fried) when it is going to be eaten. This process gives more control and better results in the professional restaurant kitchen.

COLD COOKING: BRINING, CURING, SMOKING

In the beginning of this chapter, preservation was mentioned as one of the principal reasons for food preparation. So far, a lot of focus has been given to all kinds of cooking techniques that involve heat in some way or other. This may have lead to the impression that we indeed need a fire to make our food palatable. This is not the case and, therefore, it is good to pay attention to what can be summarized as cold cooking—where acids, salts, air, and enzymes do the work of denaturing and partly digesting proteins and other large molecules. In the process, new flavors are created. Most of these techniques were originally developed for preserving food, which enabled foods to be eaten throughout the year, out of their season.

The first techniques to mention are brining or curing. The words are often mingled. Their common denominator is salt. Brining is done with water that is saturated with salt and usually some sugar. Curing is done with salt and other substances, especially nitrates. The sugar may be added for flavor, but the salt is used functionally. Salt is hygroscopic and attracts water. The texture of the product that is brined or cured changes as water is pulled out by osmosis. In this case, this is desired: bacteria, molds, and yeast all need water to grow. Removing water from food inhibits the growth of microorganisms. The nitrates are also used for a reason: they help to give meat a desirable red color or rather to preserve it. Cured hams such as the Spanish *Jamón Serrano* and *Iberico* and the Italian *Prosciutto di Parma* are famous examples. Feta cheese and the Dutch herring (*maatjes haring*) illustrate that curing is not limited to meat. In fact, it is amazing to see that processes of brining/curing were already well developed in the Middle Ages—and without food scientists! Sometimes you hear people referring to these products as raw ham or herring. This is essentially wrong; the products have been processed.

Culinary history is rich and is filled with slightly altered versions of similar techniques. Brining and curing can be combined with smoking. Many hams are also smoked, just as a kippered herring or buckling, bacon, and traditional pastrami. Although it is preferably used in combination with other preservation techniques, smoking is also a technique on its own: "cold" smoking will give you, for example, smoked salmon or eel. It involves a slow cooking process at a temperature below 40°C that takes place in a special smoking cabin. Using different types of wood can give flavor effects. Phenolic compounds in wood smoke are both antioxidants—they protect the fats from becoming rancid—and antimicrobial—they slow down bacterial growth.

DRYING, PICKLING, MARINATING

A risk of curing and smoking is that a product gets too dry or dehydrated. But what is a problem for one may be a virtue for another. Dehydration is in fact an important technique for preparing foods. The reason is evident and was mentioned before: drying implies that water is removed, which hinders the growth of undesirable elements. Food can be dried by heat and cold (freeze-drying). There are many examples. Stockfish is dried cod. One of the major components of Japanese cuisine is the *dashi* made from kombu (dried seaweed, called kelp) and *katsuobushi* (dried tuna, bonito, flakes). In Chinese cuisine, but not exclusively, the use of dried mushrooms such as the shiitake is widespread. Dried shrimp is also used as a component to enrich dishes. Finally, dehydrated fruits must be mentioned. These were already appreciated centuries earlier in the Middle East; today, they are enjoyed because of their sweetness, texture, and even nutritional value.

In all of these examples, it is clear that the dried version of the original has become something else. Flavors are concentrated and new compounds are formed. Only in some cases the name changes. Plums become prunes by dehydration and, therefore, we know that prunes are different from plums and raisins are not grapes. In recipes, fresh mushrooms cannot be replaced by dried mushrooms of the same sort without realizing that quantities may need to be adapted and that flavors will be different. If proteins are involved, it is not rare that umami comes into play. Dried mushrooms, shrimp, and also Iberico ham and other cured, dried, and fermented products are rich in free glutamic acids or 5'-ribonucleotides, inosinic acid (more in fish and shellfish), and guanylic acid (more in mushrooms).

The word *fermented* has been used but has not been introduced yet. In this case, the microorganisms—yeasts, molds, and bacteria—are set to work. Yeasts are particularly known for converting sugars into alcohol and into carbonic acid. Molds and bacteria can transform acids—chiefly into lactic acid in sour foods. Common examples are yogurt, sauerkraut, dry sausages, vinegar, and pickled foods. All kinds of vegetables and fruits are known in a pickled version throughout the world. These techniques make use of either brine or vinegar and inevitably lead to foods that are salty or acidic and that are especially rich in lactic acids.

Acidity in the form of vinegar or lemon juice can also be used to denature proteins. This is the secret behind making a true carpaccio or ceviche. An acidic marinade is spread over the meat or fish and the process of denaturation sets in. Just the same, marinating is known to tenderize tough cuts of meat because connective tissues (collagen) are broken down. Some fruits contain natural enzymes that can also be used in this respect; fresh papaya contains papain, and pineapple bromelain. Both can be used as meat tenderizer. But not too much: they can make meat mushy.

VERY COLD COOKING: LIQUID NITROGEN AND OTHER MOLECULAR NOVELTIES

For many people, the use of liquid nitrogen in the modern restaurant kitchen is the hallmark of molecular cuisine. Nitrogen is an odorless, colorless, nonflammable, and nontoxic gas. It is also very abundant; about 78% of the air we breathe is nitrogen. Most gases turn into liquid when they get cold enough. In the case of nitrogen, this occurs at −196°C. This makes it interesting to use in the kitchen. The reason is heat transfer. The bigger the difference between the temperature of the food to be prepared and the source of heat, the faster the process goes. In the case of liquid nitrogen, there is an incredible difference; therefore, the transfer of heat is extremely fast. This phenomenon has enriched the culinary repertoire, enabling processes that were not possible before. Just a few drops of liquid nitrogen will flash freeze any food it touches; you can make an instant ice cream. Oil turns solid and, when cut with a knife, oil that is not yet frozen flows out. Other soft substances such as cheese become brittle as glass and break into mysterious, rocklike forms. And in the process of its use, liquid nitrogen evaporates and generates a dense fog, which is fun to see. A special effect is what is called cryosearing.[10] This involves chunks of precooked meat that are first dipped into liquid nitrogen and then deep-fried. It brings the temperature difference to an extreme of almost 400°C. The result is succulent meat on the inside with a wonderful crisped skin on the outside. There is one thing to remember with eating food that has been prepared in liquid nitrogen: wait a while before eating it. The extreme cold can give you painful cold burns; furthermore, the cold numbs your taste buds, which diminishes your tasting capacity.

Spherification is another technique that is used extensively in modern kitchens. It refers to a liquid that is encapsulated in gel spheres. The gelling reaction occurs between calcium chloride and alginate. Alginate is a hydrocolloid, a gumlike substance that is extracted from brown seaweed. If a juice or a mousse of something (a fruit or vegetable, beverage, or whatever) is blended with calcium chloride (a specific salt) and then dipped into a mixture of alginate and water, a coating gel is formed. It can also be done the other way around: the alginate is mixed with the ingredient(s) and then put into water with calcium chloride. If tiny droplets are dripped into the alginate you get flavorful pearls, euphemistically also referred to as caviar. Spherification is just one of many possibilities for using gels. There are all kinds of gels, and most of them they have different properties. Their behavior at diverse temperatures—even at higher temperatures—has produced all kinds of new applications and surprises on the plate.

The use of liquid nitrogen and gelling agents illustrates that—with the use of certain techniques, tools, and components—chemical transformations hitherto unknown in the kitchen can now be realized. They give new flavors and textures that were not known before, and they can be used to make new interpretations of "old" dishes. Therefore, it would be better to speak of "innovative cuisine" instead of molecular cuisine. Innovative cuisine is not confined to the kitchen. In what can be called three-dimensional cooking, other senses are triggered to enhance the tasting experience. Heston Blumenthal[11] developed a dish called "the sound of the sea." It consists of cuts of raw fish and some edible items that you find at and associate with the seaside. It is served with an iPod that gives you the sound of breaking waves, which brings you to the beach and enhances your perception of the flavors. Other experiments involve bringing in smells and music or changing environmental colors, and undoubtedly there is more to come. The modern cuisine is an elaborate one that requires a lot of work, knowledge, and material. It has allowed science to move in and can truly be spectacular. It also drives prices up and sometimes makes you wonder whatever was wrong with traditional methods and meals.

COOKING TIMES AND THE INFLUENCE OF ALTITUDE AND PRESSURE

Cooking times have not yet been explicitly mentioned with regard to getting or keeping flavor. Yet most people know that over- or under-cooking food is strongly related to flavor. Time is an essential part of practically all preparations, and good timing is critical for a good result. We want to capture flavor and not lose it; and at the same time, we fancy a pleasant mouthfeel that fits the food. Despite importance of cooking time, it is impossible to answer the question "what is the right cooking time?" because—besides temperature—this depends on at least two major variables: size and pressure.

Of the two, size is the most evident. Larger pieces need longer to cook. For best results, sizes need to be in line with the desired cooking time. For instance, if a piece of meat is expected to be braised for two hours, the carrots in the sauce need to be cut big enough to be ready when the meat is ready. If the same carrots are wrapped with fish to be prepared *en papillotte*, they need to be cut smaller because this dish does not need more than 30 minutes to be ready.

But that is not all: time and pressure are interrelated. Cooking times diminish at higher pressure or get longer at lower pressure. This last phenomenon is well known to everybody that has experience cooking at high altitudes, say 3,000 meters and higher. Charles Darwin observed in his *Voyage of the Beagle* that when he was at that altitude in Argentina, potatoes did not get cooked even though the water boiled. Water boils at 100°C and freezes at 0°C at sea level or rather at an atmospheric pressure of 1 bar. At high altitudes, pressure is lower, causing water to boil at a lower temperature, which basically implies that cooking times will be longer. However, the example mentioned earlier shows that just cooking longer may not always solve the problem. In some cases, a certain minimum temperature is required; and furthermore, longer cooking times can easily lead to undesired changes in texture or loss of flavors. There is also more evaporation at high altitudes. Water cannot exceed its own boiling point. If the heat is turned up, the water will boil away faster, and whatever you are cooking will dry out faster.

For raising temperature and speeding up cooking times, pressure needs to be raised, and to achieve that, special devices are required. The most familiar of those is the pressure cooker. It contains the steam that is formed within the pan, which raises the temperature to around 120°C. It can be used effectively for all kinds of preparations, from stews, soups, and sauces to fast-cooking vegetables—mainly beans—and vegetables and nuts that are destined for purees. If the Maillard reaction is desired, a pressure cooker can also do wonders for caramelizing onions or carrots or for preparing a classic sauté, for instance.

Raising pressure is not limited to cooking; pressure frying is also possible, although not in the household kitchen. The device that made it possible was commissioned by Harland Sanders, the founder of Kentucky Fried Chicken. This was before industrial farming gave us chickens that move as little as possible and consequently provide us with tender meat that is easy to prepare. In Sanders' day, chicken legs had dark meat with a lot of collagen. This is the way nature made them, and preparing them was a challenge for Sanders because he also wanted short cooking times. Again, simply turning up the heat would not solve the problem. The patented deep-fryer that solved the problem uses high pressure rather than high temperature. It is called the Collectramatic.

A MICROWAVE IS NOT AN OVEN

Many households have one and it is often called an oven, but there are many misconceptions about a microwave. The technique of cooking with microwaves is totally different from conventional ovens. In conventional ovens, infrared radiation is used that cannot be seen by humans but can be felt as heat. The wavelength of this radiation ranges from 780–106 nanometers (10^6 nm = 1 mm). In the microwave, radiation with a wavelength of 12 cm is used. These waves, reflected over and again by the walls of the microwave, will cause tumultuous movement in the food's water molecules due to their dipole character. These moving water molecules cause an increase in temperature. Therefore, the radiation of the microwaves has nothing to do with nuclear radiation; the wavelength of gamma radiation is around 0.01 nm and of roentgen radiation around 1 nm.

Microwaves are invisible light waves that are found between common radio and infrared frequencies. This is the same group of waves located where radar, Wi-Fi, and Bluetooth are found. Microwaves are tuned to a frequency that causes water molecules and fats to rotate. These types of molecules are polar—with a positive charge on one side and a negative charge on the other. In an alternating electromagnetic field, the molecules start spinning in a constant effort to align themselves. Kinetic energy is produced between the molecules in a way that is comparable to rubbing one's hands together. In fact, the food is heated by conduction, just like boiling, and certainly not by convection as in a "real" oven. Therefore, microwaves are not warm; dishes or bowls are only warmed by the food that they contain.

Understanding what happens with microwave cooking helps to put an end to other misconceptions. For instance: microwaves do not heat from the inside out, they penetrate only about two centimeters. This means that larger items are not really suited to being heated using microwaves. On the other hand, small items take a long time to cook in a microwave. Microwaves are long—so long that they often miss small food items. Microwaves work at their best with water molecules that are liquid: the molecules start spinning easily. Frozen liquids take a lot longer to warm up.

It is also important to realize that Maillard-like reactions and caramelization will not take place—how could they ever develop? Furthermore, some textures change by microwave cooking. Pastry dough becomes tough and rubbery, and not all sauces or soups can successfully be heated by microwaves. If they are thickened in a way that does not support boiling temperatures, they will fall apart.

Not everything is bad about using microwaves, they just serve different purposes. In fact, they can be used quite well for several purposes. Examples are:

- Cooking green leafy vegetables such as cabbage and broccoli with just a bit of water and covered with a plastic wrap

- Cooking root vegetables and bulbs that are vacuum sealed, such as carrot, potato, turnip, artichoke, fennel, and leek
- Making crisp leaves of herbs such as parsley, basil, and chervil by placing them individually on a plastic wrap with a touch of oil
- Making powder of vegetable purees such as from tomato, beet, and carrot

The last two examples are rather easy and interesting applications that can be used to give a special touch to a dish or to its presentation.

MAKING THE DISH COMPLETE: SAUCES, HERBS AND SPICES, VEGETABLES, FRUITS, AND MUSHROOMS

Our focus shifts from the main ingredient to additions that make the dish complete. The role of these add-ons is not to be under-estimated. From a health and an environmental perspective, it is suggested that we should eat more vegetables and fruits. They contain vital nutrients that our body needs. Even though the specific functions of these nutrients may not always be known, there is no doubt that a healthy diet should contain enough fruits and vegetables.[12]

As a reason (or should we say excuse?) not to eat enough of them, people often mention that we have a tendency to dislike their flavor, which may even have a genetic explanation. Indeed, many vegetables and fruits originally contained toxic elements and tasted quite bitter. Through selection and advanced techniques, these negative elements have been reduced. Furthermore, preparation, sauces, and herbs are there to help. Flavor problems can be solved, and food can be made attractive.[13] With sauces, herbs and spices, vegetables, and other garnishes, dishes attain their final character and appeal. These accompaniments are often the ones that make or break the combination with the beverages that are drunk with the meal. Therefore, there are plenty of reasons to look closely at them. Let us start with the sauces.

THE ROLE AND TYPES OF SAUCES

Traditionally, sauces have a remarkable reputation in the world of gastronomy. The reputation of a chef used to be based on the quality of his sauces. These days, the role of sauces seems to be less important. An explanation could be that with an improved understanding of cooking techniques, moisture in main ingredients is better retained. Consequently, there would be less need for a good sauce that replenishes what has been lost. Fortunately, sauces have other functions as well: they can give depth and richness to dishes and can concentrate and accentuate flavors. Interestingly the word *sauce* has the same root as salt. Indeed, true sauces are flavor enrichers. Sauces are important for mouthfeel. In making sauces, two elements of sauces are critical: capturing flavor and the way fluids are thickened.

Both essentials come together in the mother of many sauces: the stock. It starts off with water, to which aromatics (herbs, spices, vegetables, chunks of meat or fish, bones) are added. They gradually dissolve and concentrate, and after five to seven hours, the sauce is born. When stock is further reduced, it ultimately turns into what is called a demiglace. It is gelatinous and full of flavor, rich in umami. The viscosity of the stock is not only dependent on time (i.e., the level of concentration) but also on the ingredients that are used. Roasting the ingredients will give extra aromas. A stock of fish will never be very gelatinous and a stock of beef will be less viscous if meat is used with little collagen. Collagen provides the gelatin that gives body to the sauce, but it takes time to extract it. Bones of young animals are particularly rich in connective tissue and, therefore, in collagen. For a gelatinous broth, calf bones are the best to use. Whatever the basis, stocks can be used to make all kinds of sauces. Good stocks do not contain fat or added salt.

Making a stock is a time-consuming process, and alternative methods of concentration have been developed, such as freeze-drying, osmotic concentration, and cryoconcentration. Such techniques are based on getting water out of the mixture in other ways than evaporation. Although this leads to acceptable results, the sauces are not the same as the classic version; they are lighter, less gelatinous, and less rich in umami compounds. However, other aromas are more obvious; they have not evaporated.

Using extraction and concentration is just one way to thicken liquids. Starch is another. It is the most familiar of thickeners. The classic example is called roux. Equal parts of fat, butter, or oil and wheat flour are heated in a pan. It is basically ready when the flavor of raw flour has disappeared. However, it can also be cooked longer to attain a certain desired color. The three basic sauces of the traditional French cuisine, béchamel, velouté, and *espagnole* are all made with roux; other sauces originate from them by adding a liquid such as a bouillon. This works best when there is a good difference in temperature: when the roux is still hot, the liquid needs to be cold—or the other way around.

The disadvantage of using starch is that it takes away some of the flavor. That is one of the reasons that today other types of thickening agents are used, hydrocolloids. They dissolve easily in warm substances and have no negative effect on flavor. The most prominent of them is xanthan gum. It can handle almost every thickening task.

Quite a different category of sauces gets its texture by emulsion. This is a mixture of two or more liquids that normally would not blend together, or only for a short time. Oil and water normally do not mix, but this is a mixture where the two "enemies" are kept together by something that is called an emulsifier. The most familiar of all is egg yolk. It is the agent that coats oil droplets in such a way that they no longer can bind together, giving space for water molecules and resulting in beautiful sauces such as mayonnaise, hollandaise, cream sauces, and zabaglione. Emulsions are technically unstable; eventually the mixture will fall apart. In the case of a vinaigrette, this happens quickly; in a mayonnaise, this may take years. While making sauces, the emulsion sometimes suddenly falls apart. This is called curdling. In warm sauces, this can be caused by overheating; in cold sauces, the most probable cause is too much oil. A curdled sauce has lost its texture and, therefore, much of its appeal and function.

Energy is needed to produce an emulsion. Whisking needs to be done either by hand or mechanically—in one of many types of food processors or blenders. This brings us to the last aspect of sauces: air. Mixtures can be whisked to become foams, and sauces may also be aired with an immersion blender. This makes sauces lighter, but the inevitable consequence is that the original sauce needs to have quite some flavor intensity to start with. The sauce is dispersed with a lot of air bubbles that are flavorless.

To finish this section: sauces can have a big influence on wines. Although they can be potential hazards, I prefer to see sauces as ideal agents to form bridges with wines or other beverages. Sauces can be the secret tool of the gastronome to produce perfect combinations. Adding just a little bit of the wine that will be served with the meal to the sauce often does wonders.

THE ROLE OF HERBS AND SPICES

In the flavor profile, herbs and spices mainly influence the flavor type. Many green herbs bring the flavor profile to "fresh." Think of parsley, chervil, dill, mint, tarragon, anise, chives, sorrel, and also of watercress, fenugreek, fennel, star anise, ginger, horseradish, mustard, and sumac. If ripe flavor tones are desired, make a choice of thyme, rosemary, garlic, bay leaf, oregano, or spices such as vanilla, nutmeg, cinnamon, cardamom, caraway, cumin, and cloves. Some herbs and spices are not very specific in their contribution to the flavor type. Well-known examples are basil and sage. Those herbs have both fresh and ripe notes.

The important question with regard to herbs and spices is when to add them. The aromas of especially the green, fresh ones are very volatile and best added just before serving. Fresh-cut parsley, chervil, or dill are just a few examples of green herbs that are often sprinkled over vegetables such as cauliflower or potatoes or over a broth. Other aromas need time to develop and dissolve in a cooking process. This is true for bay leaves and many spices.

Curry is not a spice but a mélange, and the word is also used to indicate a class of dishes. In Indian cuisines, a dish made with a mixture of many spices that are roasted before adding the main ingredient (meat, vegetables) is designated as a curry. Every individual curry requires its own mixture of spices, and ideally, the mixture should be freshly prepared. A freshly made curry mixture is always more pungent than a mixture that has been kept for a while. In colonial times, the English developed a kind of standard mixture they designated as "curry powder" that was supposed to be suitable for all dishes. It is not, but it can be excellent in itself. Curry powder is a mixture of chili peppers, ginger, fenugreek, caraway seed, coriander, cloves, and turmeric (which gives it its yellow color). The different spices are toasted dry and then ground up, or crushed. Curry mélanges vary from mild to very spicy, depending on the sort and proportion of spices used; different types of curry might give quite different effects on the flavor profile.

Creating mélanges of herbs and spices is an art that goes way beyond the world of curries. If done well, the mix can give interesting twists to dishes. Just as with sauces, such mixes can also be quite influential in combination with beverages.

VEGETABLES

Vegetables can be used as a dish in their own right. For example, asparagus is eaten in quantity with just a little ham and egg and some melted butter. In that case, the animal foods are the garnish, an addition to the vegetable. Most of the time, it is the other way around: the vegetables are chosen to enhance the flavor of the meat or the fish. Generally, the quantities of vegetables that we eat can be quite substantial, especially in home cooking.

Vegetables can supply a large variation of flavor components. This is partly influenced by which plant family the vegetable belongs to. Botanists have classified plants into different families. The initial and most important criterion was the structure of the flowers. Interestingly, plants that are placed in a particular family because of their flower structure often contain similar flavor-giving components. In other words: their evolutionary relationship is not only reflected in the structure of the flowers and the plant itself but also in specific chemical components.

Originally, plants evolved these flavors to deter animals from eating their leaves, roots, and unripe fruits. Humans have gradually selected tender and mild-tasting varieties of these plants, and now these are known as vegetables. Try to eat a leaf of wild lettuce or dandelion; it is not very pleasant. On the contrary, it is extremely tough, hard, and bitter. Yet cultivated forms of these plants are more tender and only mildly bitter in a way that people like.

We will discuss vegetables from the following plant families:

* The onion family (Alliaceae, or the genus Allium within the Liliaceae). All plants of the onion family share one typical aroma that becomes perceptible as soon as the plant is damaged; it is a sulfur-containing substance called allicin. It can be found, with small variations, in vegetables such as onion, shallot, leek, garlic, and chives. Allicin disappears if it is heated. Therefore, the use of members of this family is quite dependent on their preparation.
* The cabbage family (Cruciferae, Brassicaceae). The cabbage family also contains a variety of sulfur-containing substances, also liberated when the plant is damaged: mustard oils. These can be very aggressive and highly concentrated, such as in mustard and even more so in horseradish, but they also can be fairly mild, such as in raw cabbage, cauliflower, and Brussels sprouts. In these latter vegetables, cooking will result in the formation of trisulfides that are the source of the odor of overcooked cabbage that many people do not like. In many plants of the cabbage group that are preferably eaten raw (all kinds of cress, radish, rucola), the mustard oils are fairly mild.
* The goosefoot family and some related families (Chenopodiaceae, Polygonaceae, Portulacaceae). Spinach, beet greens, and spinach beet belong to the goosefoot family. Sorrel, rhubarb, and purslane belong to different, although related, families. They share in the first place a high content of oxalic acid (explaining the feeling of a deposit upon the teeth) and also some earthy tastes, summarized under the name of geosmins.
* The bean family (Fabaceae, Papilionaceae). The bean family includes many edible plants. Humans can eat the seeds of the plant, either as fresh, unripe seeds or as ripe seeds, first dried and soaked before cooking. Alternatively, one can eat the unripe fruits of certain varieties (e.g., green beans and mange-tout peas). All members of the bean family contain poisonous proteins called hemagglutinins. These can cause agglutination of the blood cells. Different species possess these substances to different degrees. Sufficient cooking should be the rule for all members of this family. The taste of the fresh seeds should be slightly sweet with vegetal tones.

Dried, soaked, and cooked seeds ideally have a creamy structure thanks to well-cooked starch with slight umami undertones. In the fresh fruits, a vegetal taste combined with a certain level of sweetness is desirable.

- *The carrot family (Umbelliferae, Apiaceae).* Vegetables such as carrots, celery (including celeriac), fennel, chervil, and parsley, and also herbs and spices such as cumin seed, aniseed, coriander, and caraway belong to the carrot family. Their closely related aromas mostly fall into the phenolic category, such as carvacrol.

- *The nightshade family (Solanaceae).* The nightshade family is one of the most fascinating ones. Most plants contain bitter tasting poisons that belong to the chemical group of alkaloids. Surprisingly, in a number of species, these alkaloids are missing in certain parts of the plant: in the tubers of the potato and in the fruits of the tomato, the aubergine (eggplant), the sweet peppers, and the different chili peppers. These vegetables exhibit a great variety in tastes. The peppers are extremely outspoken, due to the presence of capsaicin, a substance that causes a pleasurable irritation in the mouth (pungency, hotness). The tomato provides an acidic freshness, combined with some sweetness and above all umami components. The eggplant has slight bitter tones and low flavor intensity. The qualities of this vegetable are best brought out by severe heating (grilling, frying in oil). The potato, finally, has developed into the first or second most important staple food in Western Europe and into an important vegetable elsewhere—probably because you can do so much with it.

- *The cucumber family (Cucurbitaceae).* The cucumber family consists of herbaceous plants rambling or creeping over the ground, with extremely long stalks. The edible fruits range in size from big (cucumbers, gherkins) to huge (pumpkins). All these fruits have low flavor intensity. Cucumbers are fresh with a slight acidic and occasionally bitter taste. Pumpkins are somewhat mealy in structure at their worst and creamy at their best. They must be cooked or grilled for best results; they can be quite agreeable, especially when Maillard components can be formed.

- *The lettuce family (Compositae, Asteraceae).* The salad leaves belong to the *Compositae* family. A general characteristic is that the family members such as lettuce, endives, escarole, batavia, edible forms of the dandelion, and radicchio rosso possess bitter components that often are contained in a milky juice flowing from the plants when they are cut. The root vegetables from this family often possess, instead of starch, a substance called inulin (a polymer of fructose) as their food reserve.

The plant family is not the only factor contributing to the flavor of vegetables. Also the part of the plant that is consumed plays a role. A root vegetable will contain starch or inulin and thus be more coating and less fresh than a leaf vegetable from the same family. Sometimes one can observe this in different forms of one plant. Celery (*Apium graveolens*) is found as a stalk tuber, called celeriac; as a form with thick, tender leaf stalks, sometimes known as foot celery; and finally, the leaf celery, closest to the wild form. The celeriac in its cooked form will be creamy and coating whereas foot celery is less coating and fresher.

Finally, the way of preparation plays an important role. In the section "Characteristics of the Principal Techniques" in this chapter, the effects of different preparation methods have been reviewed; for vegetables, just one example will do. Celeriac can be used, grated or finely cut, in a salad; then it has fresh and spicy tones and an affinity with acids. As stated before, when cooked, it is rather coating and has affinities with ripe and umami tastes.

FRUITS

Fruits are a special category in the edible plants. Here, the distinction between the botanical and the culinary concept of "fruit" must get some attention. A botanist calls everything a fruit that arose from a flower and that contains seeds—one or more. In this sense, a grain of wheat is a fruit just as a tomato or a green bean. But the latter two are called vegetables in the culinary and commercial world, and the wheat is called a cereal. The term *fruit* is reserved for those (botanical) fruits that can be eaten without any preparation and possess in their flavor a delicate balance among sugars, acid, and aromatic—mostly water-soluble—components.

The use of fruit in the kitchen has a long tradition. Think of wild duck with orange or cherries and steak Hawaii. Melon and ham are often eaten together, and venison is often served with a halved apple with cranberries; applesauce or cooking pears simmered in red wine are popular garnishes, just as are other compotes, mango chutney, and Cumberland sauce. This use is not limited to traditional French cuisine. In Thai, Chinese, Arabic, Indian, and African kitchens, there are many examples. Spiciness can be softened with some fruits. It is not unusual at all to use fruits in a variety of dishes. What is the general role of fruit in the kitchen of today? We distinguish three types of influence:

• The influence of sweetness
• The influence of acidity
• Aromatic influence

The influence of sweetness is rather clear: most fruits are sweet to some degree, and adding fruits to dishes adds sweetness to them. The choice of a certain fruit in the right quantity can be functional. We know by now that sweetness can mask bitterness or soften acidity and spiciness. An example of this use is to cook some apples together with red cabbage or sauerkraut.

The influence of acidity is much more complex, and this has partly to do with a change in the general philosophy of cooking and the use of ingredients. Until the 1970s, it was quite normal to use canned fruits in the kitchen. The slice of pineapple on a pizza or steak Hawaii was never fresh, and neither were the grapes that were used for the sole Véronique. The canning process takes place at a high temperature and changes the characteristics of the fruit. Fresh malic and citric acids are transformed into milder acids, texture changes, and the sugar syrup of canned fruits is usually sweeter than the fruits in their pure form. These days the use of cans is frowned upon in most contemporary kitchens. A fresh pineapple or grape can hardly be compared to its canned alternative, and the fresh and canned versions are not interchangeable without consequence. Note that the flavor of fresh fruits also changes in the cooking process: textures change and aromas are prone to evaporation.

Aromas are the third reason why the use of fruit in dishes can be very interesting. Many fruits have a characteristic aroma that may enhance certain other flavors. "Catching" these aromas is a challenge. They are volatile and can easily disappear. Furthermore, some aromas are soluble in water, although others are dispersed in fat. This can be experienced with citrus fruit. The aromas of the juice are different from those of the peel. The original recipe of the classic orange-flavored duck requires some orange peel to simmer with the sauce. Likewise, the peel of orange or lemon can also be used to flavor a cake or a salad.

All in all, there is every reason to be positive about the use of fruit in sweet as well as in savory dishes. They can supply fresh, acidic, sweet, and aromatic elements that enhance other flavors or complete a dish. All kinds of ingredients may become more interesting by using fruits wisely. And, in their own right, fruits are used to make a huge variety of desserts.

MUSHROOMS

Superficially, mushrooms resemble plants—they cannot move about, and they have cell walls. The big difference is chlorophyll. Mushrooms are unable to use light for making carbohydrates. For their energy, they are, just like animals, dependent upon the carbohydrates produced by plants. The composition of their cell wall is different from plants: it is not cellulose but a carbohydrate containing nitrogen as well. This substance is called chitin, and this is also part of the outer skeleton of insects. This partly explains the meat-like quality of many mushrooms. Apart from that, they are per unit of weight much richer in protein than most plants. In this way, they can contribute the umami effect to food. Furthermore, mushrooms contain certain minerals—particularly zinc. The other important source for zinc in the human diet is meat, which implies that vegetarians especially should watch their intake of this element and include mushrooms in their diet. Remember that this element appears to be important in the regeneration of cells. Zinc deficiency reportedly leads to disturbances in flavor registration caused by a decrease of taste buds per papilla.[14]

SEASONALITY OF VEGETABLES AND FRUITS

These days, vegetables and fruits are available year round, either from heated greenhouses or they are imported from far away. In many parts of the world, seasonal availability of foods is fading. This situation has severe environmental disadvantages, as these foods are not always as attractive as fresh, seasonal vegetables. Furthermore, there seems to be a tendency to take sustainability seriously by looking at the "carbon footprint" and refocusing on local production. When planning to cook with local foods, it is good to know about the availability of fresh vegetables and fruits that are either open-grown or grown from cold frames or in unheated greenhouses. Clearly, their availability is strongly determined by the region and its latitude and, consequently, by the season. In fact, seasons are especially an issue between around the fortieth and the fifty-fifth latitude.

In this section, we will elaborate on the nature of plants from a seasonal perspective. The most important factors are:

- *Which part of the plant is eaten?* This is important to know because leaves come before flowers, fruits, or tubers can be formed.
- *Where does the plant come from?* This is important to know because of the temperature requirements of the plants and the requirements as to the day length.
- *Is it an annual, a biennial, or a perennial?* This is important to know because the formation of big buds or other forms of food reserve determines the availability of the edible parts.

In early spring, we first have vegetables that will germinate at relatively low temperatures, that will grow fast, and of which we eat the young leaves. Examples are spinach, spring greens, chervil, and rocket (rucola). Formation of just a few leaves takes only a very short time. All these vegetables are annuals: they perform their natural life cycle within one year.

In the second place, there are perennials such as sorrel, chives, rhubarb, seakale, hop shoots, and asparagus that produce young leaves or stalks at an amazing rate from underground food stores. Perennials can live for several years (e.g., a bed of asparagus can be productive for as long as 10 years).

In the third place, biennials are available that, with a little help from cold frames and so on, form a substantial plant before winter, survive winter, and produce their heads of cauliflower, oxheart cabbage, or lettuce in early spring. These plants originate from northern temperate areas and will only form heads (which must be seen as a preparation for flowering) if the length of the days increases. Normally, biennials grow up from seed in their first year and form substantial food reserves in late summer and autumn, from which they eventually will flower in spring, form seeds, and die.

Finally, in early spring, typical winter vegetables are still available, such as curly kale, leeks, beetroot, scorzonera, celeriac, and big carrots; but all these vegetables are past their prime.

In late spring or early summer, the young leaf vegetables are followed by vegetables that consist of (parts of) further developed plants. First there are lettuce, cauliflower, and oxheart cabbage, sown early in that same year. Second, the unripe seeds from broad beans and varieties of garden peas, including the mange-tout peas, can be enjoyed from the middle of May. These plants belong to the bean family and originate from the northern temperate regions. Also, tuber-forming plants that are quick growers and can tolerate low temperatures will be available: radish, small beetroot, early kohlrabi, rape, the earliest potatoes, and very early carrots. Very early carrots can only be grown using a cold frame; very early potatoes might need occasional protection against late frosts. At the same time, by the end of June, we lose the asparagus and the rhubarb, but not because they do not grow anymore. If we would continue harvesting, we would prevent the formation of underground food storage, thus endangering next spring's harvest.

High summer is the heyday of the heat-loving South American beans: different varieties of green beans (unripe fruits, eaten whole, cut up, and cooked) from July and unripe seeds of shelled beans (e.g., flageolets from August and dried beans from September). If properly stored, the dried beans can last for more than a year. Other heat-loving, fruit-bearing vegetables are zucchini, tomatoes, cucumbers, bell peppers, chili peppers, and aubergines (eggplants). All these vegetables, apart from zucchini, are better grown in an unheated greenhouse than in the open. Finally, from the second half of August, sweet corn becomes available. Of this form of corn, the unripe, sweet corns are eaten. In high summer, we also have other vegetables that love heat or at least need a hot start and will not bolt (i.e., start to flower). From July onward, bulb fennel, summer leeks, and endives come into season. Lettuce is still available if it is not too hot; otherwise it will bolt. Instead of spinach (for which it is too hot), we can eat spinach beet. And, apart from radishes, all the root and tuber vegetables are still available, but they are bigger and less tender. A leaf vegetable that needs very high temperatures (and grows extremely fast) is purslane. The vegetables from late summer can linger in part till approximately mid-October, depending on the weather.

In autumn, approximately after the first half of September till mid-November, the vegetables come into season that should grow really big and, therefore, need a long growing season or need shorter days and can tolerate—or even need—somewhat lower or even low temperatures.

Vegetables that should grow really big are foot celery, celeriac, big carrots, scorzonera, Jerusalem artichokes, the late types of leeks, big endives, radicchio rosso, forms of Chinese cabbage such as bok choy and pak soi, swedes, big cabbages (like red, white, and savoy cabbages), Brussels sprouts, and curly kale—and the biggest of all: pumpkins, weighing up to 100 kilograms. But for enjoyable eating, a weight of two or three kilograms is preferred for a pumpkin. Many of these vegetables tolerate some slight frost. For optimal tastiness, Brussels sprouts and kale even need slight frost, and during winter, they will survive moderate to hard frost. This they have in common with hardy varieties of leeks. The other late autumn vegetables should be harvested before the first serious frosts and should be stored indoors in a cool and dark place.

Fast-growing leaf vegetables that are cold-resistant can be enjoyed: lamb lettuce and winter purslane (that also need shortening days), and spinach that needs low temperatures.

A very special vegetable for late autumn and much of the winter is chicory or Belgian endive. This plant has quite a long growing season: it needs from early May until mid-October to develop a big root. Then, the leaves are cut off and the roots are replanted in earth or put in water in the dark where they will develop the well-known whitish heads.

From approximately mid-November, any growth of vegetables in the open is over. The vegetables that have been stored can be eaten during much of the winter. Directly from the field, curly kale and Brussels sprouts can be harvested. As long as the soil can be dug, in nonfreezing weather, scorzonera, Jerusalem artichoke, and winter leek can be harvested directly from the garden.

From the last days of January, the first spinach and spring greens can be sown for eating in March. And here, the story begins all over again.

FLAVOR COMPOSITION

TECHNIQUE FOLLOWS FUNCTION

The culinary world is rich and developing. Certainly these last years, with the help of science, new techniques, compounds, and equipment have moved in. Furthermore, anything can be made available. It has become clear that a chef has many possibilities to alter mouthfeel and flavor intensity. Take a chicken breast and divide it into three pieces. The first piece is going to be poached, the second fried, and the third grilled. Now let us look at the flavor profiles of the three pieces. The pure flavor of the chicken will show best in the poached version. Its flavor intensity will not be too high. In the fried version and grilled version, the flavor intensity will be higher; but it is theoretically impossible to say how much higher because it depends on the execution. What can be expected is that the ripe flavor tones will be more present. After preparation, the chicken meat itself is hopefully still juicy—or in other words, the meat has not been overcooked.

Most likely there is not only chicken on the plate; there may also be vegetables, a starchy element, a sauce, herbs, spices, and other seasonings such as salt and pepper. The dish offers an array of flavors. All of these are not there by accident; they have been put there and, hopefully, with a purpose. The flavor profile of the dish is the result of everything that is there. You need to taste it as a whole to define its flavor profile. Subsequently, this profile serves as a basis for communication to guests and possibly for selecting a certain wine or other beverage. To ensure that the people in service, including the sommelier, can deliver the best service, it is crucial that they know the flavor of the dish and can trust that the flavor profile is rather constant. This is best ensured when specific recipes are written down and obeyed once a dish has been developed and put on the menu. Professional cooking requires precision—up to the grams of salt that are added and the type of oil that is used. A change of brand may require a change in the recipe. The flavor profile must be reliable. The flavor profile of a dish can be adapted toward a specific beverage. The flavor of a wine or a beer is a given fact. The best professional chefs have an open mind and understand gastronomic experiences; they are not only focused on the perfect dish but also on the factors that influence the flavor perception. Consequently, they understand the importance of adapting recipes when circumstances change.

THE ARITHMETIC OF FLAVOR AND THE ART OF OMISSION

More than ever, chefs are aware that there is a whole new world to be discovered and that traditional practices need to be evaluated and, in some cases, even be challenged and possibly changed. The more there is, the more choices have to be made. To be able to make choices, it helps to know what you are doing. This is a fundamental difference from the traditional way of working in which you basically do what always has been done or use what is available. This section elucidates some of the general issues that may help you in doing things the right way.

The first is a psychological issue: more is not always better. It is quite natural to believe that adding things is appreciated, that they make things better. In flavor, this is not necessarily the case. In flavor, additions mostly work only to a certain extent. In terms of arithmetic: $1 + 1 + 1 + 1 = 4 + 1 = 3$. This is a type of calculating that you are not taught at school. In flavor terms: beware of the fact that a new addition may in fact be devaluating the dish. An omission can be just as vital as a contribution. Less can indeed be more. Therefore, it is often referred to as the art of omission. A good question to ask yourself when composing dishes is "what should this addition do; what is its function?" If you can answer that question, then apparently this element has a role to play and cannot be missed. Taking this a step further: in wine and food pairing, it can help to taste each individual component with the wine to see if it works or not. Any additions that have a negative influence on the wine might better be avoided.

The real, original sushi may well be the perfect example of the art of omission. A small bite that basically consists of little more than rice and a piece of seafood is deceivingly simple. Yet in the best cases, there is an astounding depth in flavor. The documentary film *Jiro Dreams of Sushi* shows what is involved to achieve this depth. The movie is about the life and work of Jiro Ono. His restaurant in Tokyo, Sukiyabashi Jiro, seats only ten, but it nevertheless has three Michelin stars. It is all about using the best ingredients, passion, and discipline. Practice makes perfect. Jiro always keeps looking for improvement even if he is already more than 85 years old.[15]

With regard to the arithmetic of flavor, the concept of synergy also needs to be mentioned: 1 + 1 = 3. Flavors can enhance each other. Adding just a touch of an herb, spice, or technique can bring out the best of a certain ingredient. This is the fascinating part of flavoring and is often mentioned as the mission of wine and food pairing. Although this is taking things too far as far as I am concerned, it is certainly true that it is important to see that flavor profiles fit. Viewing the previous paragraph, the opposite of synergy must exist as well, but interestingly there is not a real and official antonym for synergy. Antergy is mentioned, but the word *rivalry* could also be used to express what can better be avoided: having competition between ingredients that have their own value and characteristics but do not match together.

DETAILS MATTER

An African proverb says: "if you think small things don't matter, lock yourself in a dark room with a mosquito." In cooking, details are important. Many have been mentioned between the lines. Many others will come about in daily practice. Two specific elements are noted here: the roles of size and salt.

One part of culinary training is to learn how to cut vegetables quickly and evenly in cubes (called *brunoise*) or in strips (called julienne). A reason for learning this is that it looks nice when ingredients are chopped evenly. However, there is a more fundamental reason as well that has everything to do with the extraction of flavor from the ingredients and the time that they take to cook. In an extreme, this can be experienced when making coffee. Putting whole coffee beans in hot water does not work. By grinding them, we create many individual particles that all have the capacity to give flavor to the water and give us what we call coffee. Similarly, the time to make stocks and sauces can be reduced by cutting ingredients in smaller pieces.

When vegetables are used as a garnish to complete a dish, size is important as well and for two reasons. The first has been previously mentioned: cooking time. The second is more subtle. Smaller pieces are distributed more evenly and easily in a dish. If you take a forkful or a spoonful, chances are that all ingredients that the chef wanted are there. Size can influence the combination with beverages, especially with vegetables that are neutral such as peppers, onions, turnips, and tomatoes. A wine may match a dish but only when ingredients are cut to the right proportions.

Salt is hygroscopic, which means it attracts moisture. As we have seen, this is useful for conserving all kinds of foods. In the case of preparing meat, people often wonder whether to add salt before or after preparation. It is generally advised to salt after, to prevent the meat from drying out. It is easy to verify at home whether or not this actually happens. The answer is no. If you take a steak, salt it, and leave it for half an hour; there are some specks of moisture but nowhere near the amount necessary to dry the meat. To get a nice and flavorful crust, chefs salt before roasting or frying. The salt speeds up the baking and roasting processes and the Maillard reaction. When ingredients get salted right before being served, the salt has only one purpose—to increase flavor intensity.

PRESENTATION AND FOODPAIRING

The extrinsic factors of flavor have been introduced earlier, and their importance stressed. The flavor of a dish is also influenced by sight.

Presentation, form, and color are part of the tasting experience. This is the first impression, and we know that first impressions are important. But there is a tension. The flavor may intrinsically be perfect, yet the dish may not look nice enough. Something needs to be done; the composition needs to be altered. Adding or changing colors may be an easy way to achieve this. However, keep in mind what has been said earlier. An addition needs to be found that enhances the presentation without harming the synergy of flavors. A vegetable that has a "useful" color but a rather neutral flavor could be a perfect choice.

The scientific world of flavor is also in full development and is becoming more and more popular under the name "foodpairing." It is based upon the assumption that foods combine well with one another when they share major flavor components. Note that the word *flavor* is used here as the olfactory part, the odorants, the volatile aroma components. This is different from the way that we have approached flavor in this book— where it is used as the integrative concept that incorporates the gustatory, olfactory, and trigeminal pathways. Here, the volatiles in foods are analyzed with the use of gas chromatography and mass spectrometry. This analytical technique separates and identifies the various odorants. Once the key aromas of a particular food have been analyzed, they are compared to a database with the profiles of other food and beverage products. Products that resemble one another aromatically are graphically visualized using a foodpairing tree.[16] If you look for unexpected flavor combinations, you can be inspired to see products together that you would have never thought of combining. Daily practice will be more capricious than the theory suggests, mainly because flavor is more than olfaction. Hence, a similarity in key aromas may unveil a part of the mystery, but it is certainly not a guarantee of success.

CULINARY SUCCESS FACTORS[17]

PALATABILITY AND LIKING

In this book's introduction, flavor and tasting were launched as two different concepts. Although they are closely related, it is useful to make the difference and to use the words correctly in their own worlds.

The same can be said about palatability and liking. Palatability relates to flavor as liking relates to tasting. The first is product related, the second human related. The logical consequence would be that palatability could also approached objectively, just as flavor. Liking is a personal judgment and is, by definition, subjective. Both words are used to describe the positive response. The difference, however, is that palatability is in the hands of the producer, chef, winemaker, even of nature itself, while liking is the desired consumer response.

An interesting perspective opens up. Is "yummy" a coincidence? In the present chapter on the flavor in foods, it has become clear that there is a right way of doing things. It will have become more and more evident that there is a lot of objective knowledge involved. Foods that are prepared perfectly are palatable, and chances are that more people like those foods than the ones that have been prepared less perfectly, let alone those that have been ruined by a bad cook. We can also approach the question from the other side: if yummy is a coincidence, it would be reasonable to suppose that a group of people who like a certain product or combination would always be able to find a similar number of people who dislike it. We know that this is not the case. Although there are—to my knowledge—no foods that are universally liked, it is clear that there are foods that are liked more than others. Even more importantly: it is safe to assume that the ones that are well prepared are better liked than the ones that are less palatable.

This hypothesis has been tested. The first problem that needed to be encountered is: which foods to select? The most popular foods in the world are heavily marketed. The influence of advertising is hard to filter away. In the restaurant world, the phenomenon of the signature dish is well known. These are the characteristic dishes that identify a chef. They each do not start off as a signature dish but grow to become one because they are liked by customers. The customers share their enthusiasm with their friends, who then go to the restaurant and specifically ask for this certain dish. There is no publicity, except for reviews in magazines or online; the dishes attain their status by word of mouth communication. These dishes need to be intrinsically special.

Do these signature dishes have common denominators? To research this, successful dishes of eighteen of the best chefs in the Netherlands were analyzed. Six product characteristics, culinary success factors (CSFs), were found, and these were tested. This was done by having a Michelin star chef develop three series of dishes each with a specified main ingredient (turbot, prawn, and beef, respectively). Every series consisted of one dish based on the CSFs and two variants in which one of the CSFs was left out, under the condition that the dish was still restaurant worthy. These nine dishes were served to a group of experienced restaurant guests in a tasting. In the tasting, the dishes in which most CSFs were united were preferred to the variants. The conclusion was reached that formulated CSFs can help chefs in the development of new dishes and in the improvement of existing ones. The CSFs are discussed in the following sections.

FACTOR 1: NAME AND PRESENTATION FIT THE EXPECTATION

Extrinsic factors are important. Package, color, presentation, and image are examples of factors on the outside that influence expectation. The description on the menu should be precise and in line with what can be expected; no false expectations should be raised. Expensive ingredients are potential dissatisfiers, especially if they are explicitly mentioned in the description. Such ingredients are likely to be the reason for selecting the dish. Therefore, they should be in ample supply in the dish, but often they are not because of their price.

Changing the name and presentation over the years should not be done—except for minor adjustments and evident improvements. In many cases, the main ingredient is clearly visible and, in general, the presentation is pleasing and appetizing. Evidently, the flavor of the dish should always be the same. If the recipe changes drastically, the name should change as well. Changes in either the name or the flavor of a successful product can prove to be detrimental.

FACTOR 2: APPETIZING SMELL THAT FITS THE FOOD

Smell is an important driver of palatability. It is even plausible that, of the intrinsic flavor components, smell is potentially the most important contributor to pleasure and palatability. Foods are extra attractive with a pleasant smell. Palatability increases if smell and flavor fit together. On the other hand, bad smells can destroy the pleasure of eating. Chefs mention specifically the importance of attractive odors but also that off-flavors should not be present. The importance of smell in flavor is also seen in the decline of olfactory capabilities of older people, which tends to begin around 60 years of age and becomes more severe in persons above 70 years of age. The enjoyment of eating reportedly decreases, potentially leading to inadequate dietary intake. A study involving the elderly showed that when the smell of food was enhanced, natural intake increased.[18]

FACTOR 3: GOOD BALANCE IN FLAVOR COMPONENTS IN RELATION TO THE FOOD

In palatable foods, the flavor components are well balanced, never too sweet, salty, acidic, or bitter. The right amount of a certain component is food dependent. A salad is expected to be fresh and acidic, and a cake is expected to be sweet. Contrary to what is generally assumed, acidity and a certain bitterness were often mentioned as being important gustatory components. Many chefs consider bitter to be an essential element making food easier to consume and more interesting. Food products that are universally popular, such as beer, chocolate, coffee, tea, and wine, are all bitter to a certain extent.

Finding the right balance is important. The word "too" is often used in regard to balance. Too bitter or acidic is not appreciated nor is too salty or too sweet. Such preferences are personal, which leads to the suggestion that chefs should stay on the safe side, certainly with salt and sugar. It can easily be attributed to the personal level of pleasure. "Too ripe" was also mentioned as a potential disturber. To illustrate: meat that has been fried or grilled too long, bread that has been toasted too long, French fries that have been fried too long, or a wine that has been aged in a wooden cask too long—in each of these cases, ripe flavor notes prevail, while taking away the original flavors. The complexity of the flavor diminishes.

FACTOR 4: PRESENCE OF UMAMI

The results of this study confirm the importance of umami in relation to palatability. The name and the scientists that discovered umami are

Japanese, but as a flavor component umami has no specific relation to Asian cuisine. In fact, it is a new word for an old flavor component that has always been valued as important. Umami itself does not have a distinct flavor. It is described as brothy, savory, or meaty. Depth and fullness are also mentioned. Moreover, umami produces a strong after-taste. An agreeable aftertaste is an important determinant of the overall pleasantness of a meal. Umami also influences the perception of other basic flavors. Saltiness is perceived more intensely, which implies that the actual amount of salt used can be lower in combination with an umami compound such as monosodium glutamate (MSG; in European Union countries, E621). MSG reduces the perception of sourness and bitterness and has no apparent influence on the perception of sweet-ness. Preference for umami is likely to be innate—just as are sugar and fatty acids—because breast milk is rich in glutamic acid.

It was striking to see how often natural ingredients were used that are rich in glutamate acid or fermented products that acquire glu-tamate in the process. The chefs used these products intuitively. At the time of the research, many of them did not know what umami was. In Table 4.2, the ingredients of successful dishes using umami are listed.

FACTOR 5: COMBINATION OF HARD AND SOFT TEXTURES

Chefs often look for a contrast in mouthfeel—a combination of hard and soft textures. The combination of crispy or crunchy on one side and juicy, creamy, or moist on the other seems crucial in palatability. Crispy is associated with freshness. For example: within a relatively short time, the crispness is gone from the crust of freshly baked bread, roasted steak, toast, French fries, or even a salad. As soon as it is gone, these products have lost much of their appeal.[19]

As far as the soft textures are concerned, fat is an important factor. Foods owe much of their flavor to fat. Fats serve a variety of func-tions. In some cases, fat contributes a desirable mouthfeel or texture, whereas in others fat enhances the flavor. Both influences contribute to palatability and liking. For a food product or a dish to be successful though, the softness in whatever textural form apparently needs to be

TABLE 4.2
Examples of Umami Ingredients

Examples of Ingredients Contributing to Umami	
Natural Ingredients	Fermented/Aged Products
Tomato	Soy sauce
Corn	Oyster sauce
Spinach	Aged cheese (Parmesan!)
Onion	Emmental cheese
Carrot	Cured ham
Mushroom (all kinds)	Broth
Green pea	Stock
Green asparagus	*Glace de viande*
Scallop	Marmite
Alaska king crab	Trassi
Cod	
Salmon	
Chicken meat	
Milk/cream	
Eggs	

compensated by textures that have "bite," according to the professional chefs interviewed.

FACTOR 6: HIGH FLAVOR RICHNESS

The last factor that was identified was a dish's high flavor richness. Within flavor richness, flavor intensity and ripeness were the factors that characterized the flavor of palatable dishes. It is conceivable that this result is biased by our study design. Palatable, but noncomplex, dishes and flavors are not expected on the menu of exclusive restaurants. The guests of these restaurants are often in a certain frame of mind—relaxed and enjoying themselves—ideal circumstances to savor complexity. Situational aspects have an influence on preference:

flavor profiles that are fresh and slightly acidic fit warm weather. The perception of complexity is also likely to be situational, as it requires one to be relaxed. The atmosphere in good restaurants is therefore an important ingredient in the overall enjoyment of the food.

PALATABILITY OBJECTIFIED

The CSFs of flavor are an attempt to objectify palatability. It is important to note that the CSFs do not dictate how a food product should be made, nor do they impair creativity. The chef that was instructed to compose dishes based on the CSFs had no problem in doing so; they served per our guidelines. CSFs make flavor tangible and are useful in modifying, or rather improving, existing products or formulating new ones. Up to now, the fundamentals of flavor composition have not been formulated and the search for it may well have been neglected. Without a solid backbone, cooking and food product development can easily fall into "cook and look." The failure rate of many industrial food products is high. The absence of knowing the fundamentals of palatability may well be part of the explanation.

Food quality can be considered the most well-defined and ill-defined concept in the food industry today. Food scientists or professional chefs are likely to define food quality from a product point of view. Their definition does not necessarily correspond with consumer opinion. For commercial food products, it is essential that a product is viewed as having high quality from a consumer point of view. Earlier we noted that palatability is by definition related to food products and that it is a strong driver of liking, which is defined as the affective consumer response. For commercial food companies, restaurants included, it is essential to know the "drivers" of product acceptance.[20]

There may be an interesting analogy with architecture, art, music, and biology. In all of these areas, the mathematical rule of the golden section has been proven to apply. It is the natural order with its harmonious proportions. Long ago, these were called the *proportio divina*, divine proportions. Some architects (Le Corbusier), painters (Leonardo da Vinci), and sculptors (Michelangelo) actively used the golden section in their compositions. It has also been found to apply in the harmonious structure of musical compositions (Bach and Mozart), in the ratios of harmonious sound frequencies, in the dimensions of the human body, and in the structure of plants. Palatability in taste is comparable to beauty in art. The ancient Greeks used the same word, *techne*, for both art and technique. Art was defined as the right way of making things. In all art, technique is essential. There may be technique without art, but there is no art without technique. In conclusion, formulating CSFs can be seen as a first step in obtaining a better understanding of flavor and the components that drive liking. The results show that the palatability of food is not a coincidence. It is the predictable outcome when the CSFs of food are present. The formulated CSFs will help chefs in the development of new dishes and in the improvement of existing ones. An interesting prospect for future research is to verify if these factors apply to industrially produced foods and in other cultures.

SUMMARY

Cooking is what defines us as human, and it also started our dependency on external energy sources. Culinary history is rich and has brought us to where we are now. In the last decade or two, developments in cooking have picked up pace because science has taken an interest in the culinary world. Science has provided new and better insights into traditional products, techniques, and methods. These provide us with the nutrients we need to survive and the pleasure that flavor gives us. Furthermore, food preparation is very functional because the human digestive system is not very efficient in processing raw foods.

Cooking starts with the availability and selection of the right ingredients. Agriculture in all its appearances is a human invention. Through careful selection of species and developing methods to cultivate them as efficiently as possible, we have succeeded to some degree in making nature what we want and in producing what we need and like. The affluent world has choices to make—gastronomic challenges.

The basic qualities of ingredients are the starting point for the chef. You cannot produce something if there is nothing there to start with. The characteristics of ingredients are influenced by climate and location; cultivars, varietals, and breeds; production methods; and ripeness. Every one of these factors or a combination of them has an impact on flavor and on what can be done with it.

Besides choosing the right ingredients, cooking is all about controlling time and temperature. We need to understand the essence of the techniques and their effects. Techniques such as frying, grilling, and roasting involve high temperatures and little or no moisture. These are suited for ingredients that require just a short preparation. The Maillard reaction often creates a crispy crust with positive effects on flavor. Techniques such as braising, stewing, and poaching involve lower temperatures and moisture, such as a sauce or a stock. These techniques are particularly suited to preserve aromas of certain ingredients and to transform tough connective muscle tissue into gelatin, which makes the meat tender and succulent.

Vegetables, herbs, and garnishes have a significance of their own. They have specific qualities that can constitute or complete a dish. They add flavor and components such as vitamins, minerals, natural sugars, and bitters that may complete the flavor profile of a dish. Ingredients such as fennel and cucumber may specifically have an influence on fresh flavor type. Fresh and ripe flavor types can rather easily be moved in a preferred direction by using herbs and spices.

Mouthfeel is not only influenced by the correct application of a certain technique and surveillance of time but also by adding a sauce. The way sauces are bound influences the combination with beverages. Details are important in meal composition. The world of flavor uses a strange type of arithmetic where omissions may be just as vital as contributions.

Applying certain cooking techniques, as well as adding vegetables, a sauce, herbs, and other seasonings, are all tools to create a certain flavor profile. The most palatable dishes are characterized by the culinary success factors: expectation, smell, balance, umami, variety in mouthfeel, richness in flavor, and ripeness in flavor tones. Understanding flavor in the kitchen is indispensable for preparing well-liked foods.

APPENDIX: INFLUENCE OF PREPARATION

Whitefish (for instance turbot)

- Boiled/Poached: Respects the pure flavor of the fish; mouthfeel: rather neutral (if the fish was cooked correctly); flavor intensity: low to medium, neutral in flavor type.
- Fried: Mouthfeel will be a little more contracting; more flavor richness, with ripe flavor tones. The flavor of frying should not dominate the flavor of the fish.
- Grilled: Mouthfeel will be more contracting, and flavor richness will be higher again, as will the ripe flavor tones. The flavor of grilling risks dominating the flavor of the fish.

Red Meat

- Carpaccio (thin slices of raw meat, marinated in oil/vinaigrette): Flavor intensity is on the low side and quite dependent on the vinaigrette and the garnish. The use of Parmesan cheese or roasted pine nuts is more influential than often thought.
- Pot-au-feu: The meat is cooked in the stock and has not been roasted, so there is no Maillard effect or a crust. Flavor intensity will be medium, depending on seasoning, herbs, and spices.

- Roasted: This preparation is close to what people expect with red meat. If prepared perfectly, there is a flavorful crust and a juicy inside. Flavor intensity is above average with ripe flavor tones.

Potato

- Cooked: Mouthfeel is somewhat dry, absorbing. Flavor intensity is low to medium. Varieties of potatoes will show differences. Some keep their texture after boiling, whereas others fall apart. It is good to be aware of these differences and to select the right potato for the purpose. Also, the texture of the potato variety appears to have an influence on the way it is cooked. When you compare several varieties that are boiled with the same amount of salt, you are likely to experience differences in saltiness.
- Puree: Mouthfeel will be coating, especially when butter, olive oil, or eggs are added. Such additions also influence flavor intensity, which gets higher.
- French fried potatoes: Mouthfeel is contracting because of the crust. Flavor intensity is high, with ripe flavor tones.

Bell Peppers

- Raw: Mouthfeel is contracting because of the specific acidity. The flavor tones are fresh and flavor intensity is high and rather dominating and one-dimensional.
- Grilled: Mouthfeel is notably less contracting and may even have turned to coating. This is partly dependent on the color of the pepper: the red ones are riper (harvested later) and, therefore, basically sweeter. Flavor intensity is high but now with ripe flavor tones.

Pineapple

- Fresh: Mouthfeel is on the contracting side, despite the apparent sugars. Flavor intensity is rather high and the flavor tones are fresh.

- Canned: Canning requires sterilization and, therefore, a prolonged period of high temperatures. Flavors change, partly because of the syrup that is added and partly because of transformations in the acids, which makes them softer. In short: canned fruits, including pineapples, are very different from the fresh ones. The mouthfeel is notably more coating; texture changes as well. Flavor intensity is still high but less fresh. (This applies to other canned fruits as well; think of peaches or pears.)
- Crisps: Thin slices can be dried in an oven or a microwave. Texture is again very different; the fruit has turned crisp. Mouthfeel is therefore more contracting; flavor intensity remains high, but the flavor type is more ripe.

NOTES

1. Wrangham, R. (2009). *Catching fire. How cooking made us human.* London: Profile Books.
2. Fullerton-Smith, J. (2007). *The truth about food: what you eat can change your life.* London: Bloomsbury.
3. Pollan, M. (2008). *In defense of food. An eater's manifesto.* London: Penguin Books.
4. Murphy, K. M., Campbell, K. G., Lyon, S. R., Jones, S. S. (2007). *Evidence of varietal adaptation to organic farming systems.* Field Crops Research 102:172–7.
5. Lammerts van Bueren, E. T., Jones, S. S., Tamm, L., Murphy, K. M., Myers, J. R., Leifert, C., Messmer, M. M. (2010). *The need to breed crop varieties suitable for organic farming, using wheat, tomato and broccoli as examples: a review.* NJAS—Wageningen Journal of Life Sciences. 10.1016/j.njas.2010.04.001.
6. Brillat-Savarin, J. A. (1826). *Physiologie du goût.* Paris: Sautelet. American edition: Fisher, M. F. K., trans., *The physiology of taste* (1949). New York: Knopf.
7. Brillat-Savarin, J. A. (1826). *Physiologie du goût.* Paris: Sautelet. American edition: Fisher, M. F. K., trans., *The physiology of taste* (1949). New York: Knopf.
8. Myhrvold, N., Young, C., Bilet, M. (2011). *The modernist cuisine, the art and science of cooking.* Bellevue, WA: The Cooking Lab.

9. McGee, H. (2004). *On food and cooking. The science and lore of the kitchen.* New York: Scribner.

10. Myhrvold, N., Young, C., Bilet, M. (2011). *The modernist cuisine, the art and science of cooking.* Bellevue, WA: The Cooking Lab.

11. The Fat Duck restaurant, Bray (London), UK

12. www.hsph.harvard.edu/nutritionsource/vegetables-and-fruits/

13. Wardle, J., Cooke, L. (2008). *Genetic and environmental determinants of children's food preferences.* British Journal of Nutrition 99(Suppl. 1): S15–S21.

14. Prasad, A. S. (2001). *Discovery of human zinc deficiency: impact on human health.* Nutrition 17:685–7.

15. Gelb, D. (Director). (2011). *Jiro dreams of sushi* [Motion picture]. United States: Magnolia Pictures.

16. www.foodpairing.com

17. Klosse, Peter, et al. (2004). *The formulation and evaluation of culinary success factors (CSFs) that determine the palatability of food.* Food Service Technology 4:107–15.

18. Schiffman, S. S. (2000). *Intensification of sensory properties of foods for the elderly.* J. of Nutrition 130:927S–30S.

19. Duizer, L. (2001). *A review of acoustic research for studying the sensory perception of crisp, crunchy and crackly textures.* Trends in Food Science & Technology 12:17–24.

20. Cardello, A. V. (1995). *Food quality: relativity, context and consumer expectations.* Food Quality and Preference 6:163–70.

CHAPTER 5

The Flavor of Beverages

AQUEOUS SOLUTIONS

Beverages are essentially based on water. As already mentioned, water has very useful qualities, one of them being as a solvent for a multitude of tastants and odorants. So before going into the unique aspects of tea, coffee, beer, and especially wine from a gastronomic point of view, let us discuss the base of it all: the qualities of water. We need not elaborate about its importance. More than 70% of the Earth's surface is water, and water is crucial for all known forms of life. That includes, of course, the human race. Safe drinking water is a primary concern for the existence of civilizations, and there is a strong correlation between safe water and the standard of living as expressed in the gross domestic product (GDP). The essence of water is also illustrated by its use for setting generally used standards such as temperature and weight. The Celsius scale is based upon the properties of water; and when Napoleon introduced the metric system in 1795, the gram was defined to be equal to "the absolute weight of a volume of pure water equal to a cube of one hundredth of a meter, and to the temperature of the melting ice."[1]

The words "pure water" and "melting ice" are crucial in the aforementioned definition. Apparently, water is not just water, or H_2O. As a solvent, water is capable of containing many different kinds of inorganic and organic substances. This is where gastronomy's interest in water steps in. This measure is the total dissolved solids (TDS). The substances that are dissolved in water can influence the taste of the solution in many ways. In the case of juice, the water contains, for example, sugars and acids, and its composition is influenced by the nature and ripeness of specific fruits or vegetables. The amount of sugar is an indication of the potential alcohol if the juice is set to ferment. Next to the sugar, the acidity is important—not only in quantity but also in constitution.

In the case of tap or mineral water, other substances may have an influence on the final result. You may be inclined to think "the purer the better" or that distilled water is preferred for drinking, in cooking, or in making tea or coffee. This is certainly not the case. Distilled water does have its function in batteries and chemistry but not in gastronomy. If you use it, for instance, to make coffee or tea, there will not be much flavor. Common waters contain calcium and magnesium carbonates. The more the water contains, the harder it is. Very hard water slows flavor extraction. When making coffee, hard water is not suitable. In making espresso, the crema (foam on top of a good espresso) is hardly formed, and in tea, an undesirable layer on top is formed. On the other hand, if the water is too soft, coffee and tea are overextracted and this gives a salty flavor. Soft water can also do harm to copper joints in machines such as steam ovens or coffeemakers. We can be grateful that water companies worldwide not only ensure that our water is safe to drink but also good to use in many respects.

MINERAL WATER OR SPRING WATER

In Europe, natural mineral water is defined as microbiologically wholesome water from an underground aquifer tapped via one or more natural or drilled wells (Council Directive 80/777EC of the European Union). Mineral water is characterized by constant levels and relative proportions of minerals and trace elements at the source. No minerals may be added to this water. The only treatment allowed prior to bottling is to remove unstable components such as iron and sulphides and to (re)introduce carbon dioxide. Water from a natural source that contains few minerals is called spring water. In contrast to mineral water, which has to be bottled at the source, spring water may be transported first. Mineral water has to contain a minimum of 150 mg/l of minerals. However, local standards differ per country. According to the U.S. Food and Drug Administration, water can be called mineral water if it contains not less than 250 ppm total dissolved solids.[2] In addition to microbiological criteria, there are distinctive criteria based on the chemical composition of the water.

Most countries also recognize mineral waters as having properties favorable to health: in Germany, this is called *Heilwasser*. This is the only type of water that is entirely untreated. Iron and sulphur may not be filtered out, and no carbon dioxide may be introduced. Clinical tests must have demonstrated that the water contributes to the prevention of certain complaints. An exception is made for water that has been known for ages for its wholesome effect.[3]

An underground reservoir of mineral water is called an aquifer. There are basically two types: groundwater aquifers (where water runs freely in the subsoil) and artesian aquifers. In the second type, the water is under pressure due to the geographical situation. Many artesian aquifers are reservoirs between impervious layers. Pressure builds up because the water cannot escape. The mineral content of the water is strongly related to the time that the water is in the source. This can be very long. Clearly, the properties of the rocks and earth layers, and the types of minerals that they contain, have a strong influence on the composition of the water. In general, the shorter time the water is in the aquifer, the lower the mineral content will be. The mineral content is further determined by the type of rock. Quartz, for instance, does not give off large quantities of minerals; porous volcanic soils generally produce the richest waters.

The aforementioned indicates that the flavor of mineral water can be expected to be very different. Luckily, the mineral content is stated on the label, which gives at least some indication of what can be expected.

Waters without CO_2 and with a low mineral content will generally be softer. Spa Reine (bleu) from the Ardennes in Belgium has a TDS of only 33 and is, therefore, classed as spring water. In general, the higher the mineral content, the harder the water will be. To give one example of how high it can get: Vichy Catalan from Girona in northern Spain has a TDS of 2,900. The water is rather salty because it contains 1,100 mg/l of sodium—a level that is comparable to sea water (which normally contains about 1,000 mg/l).

Some mineral waters have a natural sparkle, but many of the sparkling mineral waters are artificially carbonated these days. Some are carbonated with the gas that the water contains naturally. That gives the producers control over the amount, size, and persistence of the bubbles. More and more popular brands offer waters with different levels of carbonation.

Next to the minerals and CO_2 there is the pH, the acid/alkali balance of the water. The pH for pure water is neutral, exactly 7.0. Waters with a pH of less than 7 are slightly acidic; the alkaline waters, with a pH greater than 7, appear to taste sweeter and softer than neutral ones. Note that they actually do not contain sweet substances. It is the absence of sour and bitter that makes them seem sweet. Table 5.1 shows the mineral content of some well-known waters.

MINERAL WATER AND THE INTERACTION WITH WINE AND FOOD

If mineral water is drunk at the table in conjunction with wine, there is an extra dimension to pay attention to. Mineral waters may influence the flavor perception of wines. Table 5.1 shows huge differences in the composition of just a few popular bottled waters. These differences make it impossible to be very specific in predicting what the influences will be. The first message is, therefore, to be aware. Unwanted side effects can occur, dependent on the specific nature of the water. Nevertheless, the flavor theory brings us closer to understanding what can be expected because it parallels the factors that influence the match between wines and foods. Although this is the subject of the next chapter, we will briefly mention what may happen.

TABLE 5.1
Comparison of Some Popular Mineral Waters

	Spa Reine	Aqua Panna	San Pellegrino	Apollina-ris	Vichy Catalan
TDS	33	150	930	2770	2900
pH	6	7.9	5.4	5.8	8.3
Calcium	4.5	32	170	90	14
Chloride	5	7.2	52	130	584
Bicarbonates	15	100	210	1800	2081
Magnesium	1.3	6.6	50	120	6.4
Nitrate	1.9	0.5	0.6	–	0.06
Potassium	0.5	–	2.6	30	50.7
Sodium	3	6.6	33	470	1097
Sulphates	4	22	430	100	49.6

To start: carbonic acid and a lower pH (acidity) are contracting forces in the flavor of water. The slightly acidic waters with a pH between 5 and 6, in general, are rather harmonious with food and wines because these counterparts mostly have some acidity of their own. The flavor of some types of wine may be enhanced by mineral content or the sparkle of CO_2. This is the case with many rosé wines or wines that are drunk as an aperitif. Wines that are high in alcohol can also be enhanced by carbonic water. The water's sparkle may make the wines seem lighter and more pleasant to drink. Carbonation reduces the sweetness of wines.

The amount of carbonation and the size of the bubbles is an indication of quality, just as in sparkling wines. If they choose sparkling mineral water, wine lovers will generally prefer waters that are not too carbonated and with fine bubbles. The sparkle cleans the palate, which is also a useful role.

A common characteristic of contracting forces is that they add up. Therefore, with rather acidic food and wine, a neutral water is a better choice than an acidic one. A wine with a pleasant acidity can become too acidic due to substances from a mineral water. On the other hand, wines that lack some acidity can be helped by having a "contracting" water. Sparkling water with wines that have a sparkle of their own should be watched extra carefully.

Furthermore, the salt (sodium) content of waters is an indication of the flavor richness of the water. The higher it gets, the more dominating the water will be. Fragrant wines—and foods for that matter—are better off with water with lower sodium content. A characteristic of saltiness is that it enhances acidity as well. Therefore, sodium-rich mineral waters are not the ideal choice with acidic foods and wines.

In general, it may safely be assumed that the softer the water is, the less the interference will be. This implies that soft water is preferred when you want to enjoy the wine as it is. This is most likely certainly true with great wines with a lot of character. Examples are subtle white wines that are aged in barrels or premium red wines. They have their own intrinsic character that you do not want to change. Soft water can also be the best choice with simple, fragrant, and aromatic white wines. The fragrances could be dominated by water with a high mineral content.

There is also an influence on the flavor of water that we control ourselves: the temperature. The colder the water is served, the more contracting it will be. Sparkling water and water with a relatively low pH will be perceived as harder. The ideal serving temperature would be between 8°C and 12°C, depending on the composition of the water. Clearly, this is much warmer than the iced water so beloved by many consumers and restaurants. Carbonated water is served slightly colder than still water. Furthermore, water connoisseurs will never add ice cubes to their water because these are made from tap water.

TEA[4]

Of the brewed beverages with gastronomic appeal, tea and coffee are very close to water. Therefore, the quality of the water that is used for making tea and coffee should always be taken seriously. The ideal water has a pH that is close to 7 and has a moderate mineral content. Tap water generally has a pH above 7 and may also contain chlorine that may give off-flavors if the water is not sufficiently boiled. Hard water with a lot of minerals is to be avoided. The calcium and other salts will prevent aromas from developing.

Tea is the most widely consumed beverage in the world, after water. It is an infusion. The leaves of the tea tree (*Camellia sinensis*) give their character to the water. This evergreen plant is indigenous in Southeast Asia. China and India are important producers of tea. The foothills of the Himalaya in Darjeeling, India, are famous for the quality of the tea; but that country only became a producer after the English started drinking tea on a large scale and saw fit to use their own colonies to produce for the needs of the empire. The old name of Sri Lanka, Ceylon, lives on in the world of tea.

The leaves of the tea tree can be gathered throughout the year. In just 14 days after plucking, new shoots are formed that can be harvested again after two to three weeks. The best teas are made from the smallest, youngest new leaves of a branch: two leaves and a bud, also known as "first flush," is what tea lovers will look for. The original name is orange pekoe, where orange refers to the color of the leaves. Souchong is the name for bigger tea leaves that grow lower on the shoot. The most expensive teas are from higher altitudes where it is dryer and cooler, which causes the trees to grow slower. The bigger leaves and those from lower altitudes are less expensive.

Tea leaves are full of interesting chemicals. One of them is caffeine, especially in the *assamica* variety of the plant. Tea leaves contain even more caffeine than coffee beans. However, we need more grams to make coffee, and therefore coffee contains more caffeine than tea. Tea has a unique amino acid: theanine. It makes people relaxed and at the same time productive, alert, and focused. Tea leaves have more beneficial qualities, such as antioxidant powers from the catechins. Early man started enjoying the leaves by chewing on them, and the infusion was first primarily used as a medicine.

Infusions of fresh leaves of mint or verbena immediately give flavor, but if you would soak fresh tea leaves in boiling water, there would hardly be any change. Fresh tea leaves are mainly astringent and bitter, nature's common defense against animals. The leaves need the work of enzymes to break open cells and free, or form, aromas. Therefore, the common story that the use of tea leaves was discovered by the legendary Chinese emperor Shen Nong (also known as "the divine farmer"), when some leaves fell from a tree into boiling water, must be a myth. Nevertheless, there is no doubt that tea originated in China and also that the aforementioned ruler used teas of all sorts to develop a better understanding of their medicinal value.

Just as with any other leaf, as soon as they are removed from the tree, tea leaves gradually turn dry and eventually brown. This involves many enzymatic processes, leading to a first change of flavor. Later on, it was discovered that processes such as drying, pressing, crushing, and all sorts of heating (steaming, pan-frying, smoking) could be used to release other wonderful aromas that we now know in tea. The heat treatments ultimately stop the enzymatic processes in the leaves. The dried leaves will keep for a considerable period.

Just two ingredients, water and leaves, and the simplest of preparations, produce a huge variety of teas with very different flavor styles. They vary from light, fragrant green to heavy, imposing dark. This is the result of thousands of years of civilization and learning.

FLAVOR STYLES OF TEA

The enzymatic transformation of the tea leaves is often referred to as fermentation, but it involves no microbiological activity—it is a form of oxidation. After the leaves have been gathered, they are first withered for some time. The enzymes change the color and determine the flavor components of the final tea. Color and flavor increase if the withering is longer; the tea becomes less bitter and astringent. A combination of techniques and differences in time and quantity of pressure give the different styles of tea. The process is stopped by heating.

Every style has its own ideal infusion. Temperature of the water and the time that the leaves are allowed to give their character to the water (the steeping time) are the two main variables. Most people just use boiling water whatever the type of tea. Yet fragrant teas that have had little "fermentation," such as a green or white tea, are best brewed at lower temperatures, between 65°C and 85°C. Boiling water is suited for black teas to extract their large, complex, flavorful phenolic molecules (phenolic compounds are explained later in this chapter, when wine grapes are discussed). Steeping time may vary from 30 seconds to three minutes. The longer the leaves are in the water, the more tannic the tea will be. This may be a choice, but mostly it is advisable to limit the steeping time. The best teas can be brewed several times with the same leaves. Traditionally, in China, the first infusion is even wasted. The next infusions are drunk. The third to fifth brews are often considered to be the best, depending on the type and the quality of the tea.

We will discuss the flavor styles of green, oolong, and black teas and mention some interesting special teas. For each type, we will discuss the ideal temperature of the water for the infusion and the steeping time.

- *Green tea*. Green tea is as the color suggests and is hardly fermented. After harvest, the leaves are pan-fried in a wok, sometimes using bare hands. This is meant as an accelerated drying process. The leaves stay green. Then they are rolled to break down the tissue structure, thus releasing the cell fluids. Afterwards, they are left to dry. For Japanese green tea, the leaves are steamed, which gives them notes of grass and seaweed. Green tea is still close to the original qualities of the fresh leaf and contains just as much caffeine as black tea. The flavor intensity is not very high, yet it does contain guanylic acid (GMP) and inosinic acid (IMP) that have umami qualities. The infusion is slightly bitter because of the antioxidants, and it becomes bitterer with hotter water and longer steeping times. Ideally, green tea is brewed with water of around 70°C. The higher the quality of the leaves, the lower the temperature, with a minimum of 65°C. A practical tip is to first add enough cold water to the pot to arrive at the desired temperature when the boiling water is added.

- *Oolong tea.* Oolong is the name for the Chinese type of tea. Some modest enzymatic transformation is allowed—not as much as black tea, but clearly more than green tea. The leaves are bruised to encourage the work of the enzymes. The leaves turn slightly brown with red edges. The process is stopped after a couple of hours by heating the leaves just below 100°C. Every province of China has its own specialty; it is the type of tea where the leaves can be used several times, and the tea improves. The brewing temperature of oolong is around 80°C, and the steeping time is generally around five minutes.
- *Black tea.* For a long time, this was the tea of the West, made popular by the English. Ceylon, Darjeeling, and Assam tea are well-known examples. In these types, the enzymatic transformation is allowed to be completed. The withering takes hours; after that, the leaves are rolled for more than an hour, and then they rest for anywhere between one and four hours. By then, the leaves are reddish brown. The final stage is drying them at around 100°C, which makes the leaves turn black. In these teas, ripe flavor tones are likely to predominate. It is quite common to blend black teas of different origins to produce a certain character that can be maintained. This, of course, is an industrial interest. Black tea is made with boiling water, and the steeping time is about three minutes, again depending on the tea type. Some people like really strong tea and let the leaves steep longer. The tea will get bitterer—a problem that is traditionally solved by adding some milk.
- *Pu-erh.* The flavor of Pu-erh tea comes closest to coffee. It is a specialty of the Yunnan province in China and is the darkest of teas. In fact, this is what the Chinese call black tea. Our black tea is red tea to them, which is quite comprehensible if you compare the colors of Pu-erh and the black tea we are used to. There are several forms of Pu-erh, but they all are pressed and aged. A fermentation process takes place after the black tea is made. This tea mostly comes on the market as small compressed bricks that are individually wrapped. The flavor intensity of Pu-erh is high, especially with longer steeping times. It has very ripe flavors such as chocolate and mocha.

- *Aromatized tea.* The tradition to aromatize tea is not new. Well-known older examples are jasmine tea and tea with rose petals. Arguably the most famous is the tea named after Earl Grey. It is a black tea with bergamot. Clearly, the flavor character of the tea is influenced by what is added; the flavor type is influenced. A last example of an aromatized tea is Lapsang souchong. This tea gets its character by smoking the leaves. Like Pu-erh, it has ripe flavors and a relatively high flavor intensity.
- *Rooibos.* To conclude our discussion of styles of tea, we cite one that is not a tea in the official sense. Rooibos is not made from leaves of the tea tree but from a bush (*Aspalathus linearis*) from the *Leguminosae* family that is found in South Africa. The needlelike leaves of the bush are oxidized like tea leaves. South Africans have drunk the tea for centuries, and it has gained popularity in many other countries. One reason may be that the leaves contain no caffeine and hardly any tannins.

OTHER USES OF TEA

To conclude, it is worthwhile to mention two other uses of tea. The first is in the kitchen. Tea has many culinary possibilities. It can be used for marinating or as a poaching liquid. Evidently, the character of the tea is chosen carefully to obtain the desired effect (one that could not be obtained with, for example, stock or wine).

The second use is instead of wine at the table. This use of tea is more popular in modern Western gastronomy. This is partly due to the fact that the alcoholic content of wines has generally risen to levels that have become a burden in gastronomy. Wines with an alcohol level of 14% and more can be too imposing, too heavy. Serving tea with a course can be an interesting alternative. There are also foods that are very difficult with wine. Take some cheeses, for example. Other beverages such as tea or beer can be interesting solutions. Not just any tea—the same care that we take to select the proper wine with the dish should be taken to select the best tea to accompany the plate. Why not serve the tea in a wineglass as well—or at room temperature? After all, original iced tea has been a popular refreshment drink in many cultures.

COFFEE

Coffee and tea have many things in common. The two beverages are extremely popular throughout the world; both contain caffeine, and they are widely used in gastronomy. Many people finish their meal with one or the other, although the general preferences differ from region to region and person to person.

Both beverages are also extremely different. Tea is made from leaves. Although it is possible that the leaves of the coffee plant were also used originally to make some kind of tea, for the beverage we are now familiar with, the seeds of the cherry-like berries from the coffee plant are used. Quite some work needs to be done before we can make coffee. First, we need to get to the seed of the fruit, which can subsequently only be used after a careful process of drying and roasting. This brings us to a next big difference with tea: although coffee bean roasting is hardly ever done in the country of origin, the whole process of transforming tea leaves is done at the plantation. Let us see what needs to be done in order to get a cup of coffee on the table and see what the flavor issues are with coffee. First, we go from plant to bean and then from bean to cup.

FROM PLANT TO BEAN

The coffee plant is a tropical bush that grows all year. The most important source country is Brazil. Other important coffee nations are Vietnam, Indonesia, Colombia, and India. In total, there are more than 70 countries where coffee is cultivated. Two species of the genus *Coffea* are commercially very important: *Coffea arabica* and *Coffea canephora,* which formerly was called *Coffea robusta.* In the coffee industry, the name "robusta" has been maintained. *C. arabica* is the more interesting species, but *C. canephora* (of the robusta beans) is easier to grow. Coffee consumption started with the *C. arabica* species. It is native to eastern Africa, especially Ethiopia, Sudan, and Kenya. Today, these countries are still major producers of high quality coffee, together with Colombia and neighboring countries of western Latin America. Arabica beans contain more oil and sugar; they are the most aromatic and delicate and not as bitter as robusta. Robusta beans have more body and keep better; they also contain more caffeine. Robusta beans are generally blended with Arabica or used to make cheaper and instant coffees.

About 75% of all coffees are Arabica. Within the species *Coffea arabica*, many varietals are found, with names such as Typica, Bourbon, Caturra, Blue mountain, Maragogype, and so on. Every country and region has its specific cultivars that perform best. This is quite comparable with wine grapes, where the species *Vitis vinifera* has many cultivated varieties (cultivars) with names such as Sauvignon, Chardonnay, and Pinot noir. In the world of wine, awareness of the differences between varietals has been cultivated. In the world of coffee, this is just starting at the consumer level; clearly, the professionals have always known about the specific characteristics of the individual varieties and regions.

All the same, for making coffee we need coffee beans. Strictly speaking, it is not a bean at all but the seed of the coffee plant fruit. In tropical regions, this plant bears fruit all year, which implies that there is ripe and unripe fruit on the plant at the same time. This presents a challenge: for the best coffees only the ripe berries are used, which requires manual harvesting of the berries. Most coffee berries, however, are machine harvested; plants are stripped, with ripe and unripe fruits gathered simultaneously. After harvest, the pulp is removed either by a wet or a dry process. In the wet process, the fruit is first cleaned then "depulped" to separate the flesh from the seed. Every berry contains two seeds; each one has a "bean." To get them, the pips are first left to ferment for a while, then washed and dried in the sun or in a dryer; subsequently, the outer covering of the seed is taken away so that finally the "beans" can be harvested. Next, the beans are graded: large, medium, small (or *caracolito*). This complex washed-beans method takes up to two weeks and demands the use of a great deal of water in a region where water is not always abundant. Nevertheless, this ensures aromatic, high quality beans with good acidity. The coffees tend to have less body. The natural or dry method is faster and, therefore, cheaper; and it does not consume as much water. The berries are left to dry in the sun, sometimes after some days of fermentation. The pulp is removed by machine. The beans are not as aromatic, but they are more suitable for full-bodied coffees.

Whatever method is used, the result of the process is called green coffee; and yet no coffee can be brewed from it. Raw green beans are as hard as kernels of dried corn and have virtually no flavor. This green coffee is bought by a coffee company that does the roasting, the blending, the grinding, and of course the marketing. It is important to note that the whole process starts with a coffee company sourcing and buying the beans. Just as in wine and cooking, the origin and quality of the primary material are important in determining the final result. It is more and more customary to see origins mentioned on labels and in restaurants and especially in coffeehouses. Although this book is not about promoting one or another company, where coffee is concerned, Starbucks and Nespresso are interesting examples. The success of these companies—and the many boutique companies that may grow to be as well known—also illustrates that the modern consumer is interested in flavor and will pay the price. People do indeed taste the difference(s).

FROM BEAN TO CUP

Following sourcing are the activities coffee companies undertake to create flavor. Some consider roasting coffee beans almost as an art but, in fact, it is about mastering a series of chemical processes during which aromatics, acids, sugars, and other flavor components are changed or formed. This happens in a certain time frame ranging from 90 seconds to 15 minutes and at temperatures between 190°C and 220°C. Several stages can be distinguished. First, the green beans are slowly dried. Their color becomes yellowish and they begin to smell toasty, like popcorn. The second stage is called the crack. It occurs at approximately 205°C. At this temperature, water particles turn into steam, which makes the beans double in size. Their color changes to light brown and various chemical processes occur. Important are the Maillard reactions that give aroma to coffee. In the process of roasting, the beans become brittle and easy to grind.

Part of the art of roasting is deciding how quickly the crack should occur. After the crack, the third stage sets in. If the temperature is allowed to rise, the beans turn even browner and oil begins to escape. New aromas are being formed, but acidity and sweetness get lost. Starting at 170°C, the sugars in coffee begin to caramelize and gradually lose their sweetness. Therefore, lighter roasts are sweeter than darker ones. On the other hand, acidity rises and unpleasant bitters need a certain time at higher temperatures to degrade. At these high temperatures, the acidity changes again and becomes less apparent. Roasting is about finding the desired balance. Light roasts are more aromatic and tart. As the roast gets darker, the bitterness increases together with specific coffee aromas. Medium roasts give the fullest body. In packaging, it is crucial to find the words to communicate the roast's flavor characteristics to the customer, enabling him or her to make the right choice of coffee that fits preference or situation.

Not only the roast is important; the preferred brewing method is also strongly related to personal preference and use. The grinding of the coffee beans is related to the way the coffee is brewed and the extraction time. Filter systems lead to lighter brews and espresso is the strongest. On the other hand, the extraction time is just about 30 seconds for espresso—the briefest of all brews; filter systems take minutes. Real espresso can only be made under high pressure; in professional machines, water is pumped through finely ground beans at 9 bar. The consequence is that about three to four times as much coffee is needed compared to filter systems. The result is a coffee with much more solids, including a large amount of coffee oil, which prolongs the flavor in the sense that these oils linger in the mouth. This is a coating element and is not the only one in well-made espresso. There is also the foam on top, the crema. It is a mixture of mainly proteins, carbohydrates, and CO_2, and it is essential for the flavor of espresso. It has multiple roles in flavor: it covers the coffee and contains the aroma of the brew; it adds new flavors; and—very important from the gastronomy perspective—the crema contributes to mouthfeel. When it is absent, the espresso loses appeal and tastes much bitterer. In other brews or coffee specialties, the bitterness is masked by milk and/or sugar. Real espresso does not need milk or sugar, and the true aficionado knows that.

Every flavor world has its own specialist. In the world of coffee, this has become the barista. Originally the word meant bartender in Italian, but by now it is the one who operates the coffee machine and who knows everything about making a very good espresso or one of its derivates. Within this trade, *latte art* has developed: making an image with milk on top of the crema. It is a nice example of how extrinsic components of flavor can enhance the enjoyment of what used to be just a cup of coffee.

BEER

Beer is the last, but not the least, to mention in our gastronomic journey along brewed beverages that are dependent on water for their quality. Depending on the type of beer, around 90%–95% is water; therefore, the basic quality of water is still extremely important, just as in brewing tea or coffee. Historically, beer breweries were often founded close to sources of good water. There were times that beer was safer to drink than water because one of the steps in the brewing process involves boiling, thus killing all harmful bacteria. The German Reinheitsgebot (purity law) of 1516 is the oldest food quality regulation that is still in use to protect the consumer; it is based on ensuring that the beverage is safe to drink and is only made from clean water, malted barley, hops, and yeast.

Beer is the third most popular drink in the world (after water and tea) and the most widely consumed alcoholic beverage; it is assumed that beer is the oldest fermented alcoholic beverage. It is always interesting to look into the history of old products that are still consumed on a large scale today. There must be something special about these products. After all, the human race is good at adapting to circumstances. Products that we do not like were abandoned a long time ago. There is no doubt that early man knew about the brewing of beer—perhaps as far back as 10,000 years ago when people started growing barley, which became a necessity when people started eating bread. Relevant archaeological findings show that a brewery was often associated with a bakery. It could be that loaves of bread were used to make beer or the other way around: beer was used instead of yeast for making the early breads. Either way is technically feasible.

In any case, without knowing exactly what happened, people must have discovered how to convert starch into sugar, make some kind of sweet solution, add yeast to it, and let it ferment to end up with what we call beer. By now, science has revealed most secrets, but it remains a mystery how men have learned what they needed to do. The popularity of the brew led to a global industry that is dominated by a few huge multinational companies. Nevertheless, there are a growing number of small breweries that make local and specialty beers. All producers together, large and small, make up a beer market that is very diverse. There is much more to be discovered than the popular beers of the big brands. The biggest challenge is to get to know the differences. Strangely enough, beer types can have a similar name but be different in style.

THE PROCESS OF BREWING

Beer is, in essence, the result of the alcoholic fermentation of starchy ingredients in which the natural CO_2 remains. To make beer you need, in sequence:

- *A starch source.* In almost all cases, grain is used—especially barley because of its high starch content and because its fibrous hull remains attached to the grain during threshing. Wheat is also used. It contains less starch and more protein. Wheat gives a flavor that is slightly acidic. Other grains such as rye, corn, or rice may be used as well. A larger brewer will use grain from various sources to control quality and, possibly also, his costs.
- *Malt.* Malting is the process that transforms the grain and stimulates its starch content. This is done by first soaking the selected grain in water. After that, the grain is allowed to germinate in humid and warm conditions. This produces the enzymes (diastase) that are needed to convert the starch into sugars. When enough enzymes are formed, the germination process is stopped by drying the green malt over dry air. The malted grains can then also be roasted up to a certain level of caramelization. Different drying times and temperatures are used to produce a certain type of malt. Darker malts will ultimately result in darker beers and in different flavors. If malt is dried over smoke instead of air, this will give the malt a smoky flavor (compare to Islay malt whiskey). The malt is then crushed. This is only done at the start of the brewing process in order to prevent unwanted oxidation.

- *Water.* Water is needed to make the mixture that can be set to ferment. Traditionally, the properties of the regional water were the dominating factor in the type of beer that was brewed. The hard water of Burton on Trent produced ale, or a stout such as Guinness in Dublin. The very soft water of Pilsen gave the world pilsner urquell beer, and ultimately the most popular (in volume) of them all: lager. Most beer recipes call for the use of soft water and, if necessary, the removal of excess minerals. But for making ale or stout, the addition of water hardeners such as table salt (NaCl) and gypsum ($CaSO_4$) may be required. These days, the technique has advanced so far that any type of beer can be made anywhere in the world.
- Whatever its composition—natural or aided—the water is heated and mixed with the crushed malt. This is called mashing and takes about one to two hours. The enzymes convert the starch, and the result is the sweet tasting wort.
- *Hops and other flavorings.* The wort is vigorously boiled for about one to two hours to remove bacteria that can spoil the wort and cause the beer to go sour. Flavorings such as hops are added, which can be done somewhere during the process, at the beginning, halfway, or at the end, depending on the flavoring or the effect that the brewer seeks. Addition of hops in the beginning of the boiling process will emphasize the bittering qualities; toward the end, the aromatic properties predominate. When hops are added after the boiling process is over, this is called dry-hopping. Hops are the fruit cones from female plants of a climbing plant of the hemp family, *Humulus lupulus*. At the foot of each leaflet of a hops cone, small yellow glands are visible that secrete many different aromas, among which are the bitter substances such as humulone. They give beer its pleasant bitterness. Hops have been used on a large scale since the fourteenth century when it was discovered that the flower of the hop vine is also an effective preservative because of the mild antibacterial properties of humulone and related components.
- There are several types of hops, each with their own character and their own influence on the flavor of the beer. Besides hops, combinations of various aromatic herbs, spices, berries, and other

fruits can be used to give a specific flavor to the wort. Examples are coriander/cilantro seeds and Curacao peel (used in witbier), green tea, pepper, star anise, cardamom, licorice, or even chocolate and coffee. Small breweries are always experimenting, with wild and sometimes surprising results.

- *Yeast.* Wort is cooled down to a desired temperature range where yeasts can accomplish their part of the process. Yeast is the microorganism that is needed to convert the sugar into alcohol and carbon dioxide. Evidently, the carbon dioxide is prevented from escaping.
- A distinction is made between top-fermenting and bottom-fermenting beers. This refers to the temperature of the fermentation. The top-fermented beers ferment at what is called room temperature, more precisely 15°C–24°C. The bottom type generally ferments at a temperature between 4°C and 10°C and sometimes even lower. The fermentation process takes place for about five to seven days, after which the beer has to lager. Top-fermenting beers typically lager for two to three weeks at room temperature. Due to a slower metabolism, bottom-fermenting beers need more time—up to seven weeks at almost 0°C. This requires capital for the installation of cooling equipment—hence the development of large lager brewing beer conglomerates. Top-fermentation typically yields flowery, fruity aromatic, sometimes complex beers, whereas bottom-fermenting beers are cherished for their crisp, clear, and easy-drinking properties. Different species and even genera of yeast are required. At the higher temperatures, *Saccharomyces cerevisiae* is used. For the low temperatures, *Saccharomyces uvarum* or *Saccharomyces pastorianus* are used. To mention some other examples of specific yeasts: *Brettanomyces* ferments lambics, and *Torulaspora delbrueckii* ferments Bavarian Weissbier. Most commercial beers are filtered before bottling; however, some beers receive a secondary fermentation, either in the bottle or in cask, sometimes with a different strain of yeast or bacteria.
- More than in the world of wine, brewers like to have their "own" yeast. Different genera, species, or even strains of yeasts have specific qualities. The house yeast culture is among the most closely guarded properties of the brewer, especially in the world of lager beer where only a few other flavors are added.

BEER TYPES AND FLAVOR PROFILES

Some religious orders have made quite a contribution to the beverage world of today. Trappist monks have done that with beer. The Trappist order originates in the Cistercian monastery of La Trappe in France. The Cistercians have left their spectacular traces on wine. Trappist beers are among the most historic and best defended of all beer types. Even today, these beers must be brewed by, or at least under the supervision of, Trappist monks. In 2011, there were only seven official Trappist monasteries left, six in Belgium and one in the Netherlands. Chimay, Westvleteren, Westmalle, and Orval are among the most prestigious of beer brands.

Still, the world of beer is one of brands and labels. There are hardly any rules that determine the characteristics of certain beer types. Most are based on tradition and a broad consumer understanding. Beer is hard to officially classify. By and large, countries and breweries have their own names, and it is up to the consumer to learn the differences. To shed some light, we will mention the most important flavor instruments of the brewer; elements that can be used to make a certain type.

- *Temperature of fermentation/type of yeast.* The top- and bottom-fermented beers have already been mentioned here. We briefly mention them here again because the most popular and widely consumed beer is an example of bottom-fermentation: lager or pilsner. The word *lager* refers to the German word *lagern*, which means "to store." After fermentation, the beer is stored at 0°C–4°C. The color is light/yellow from the beginning; its appearance becomes clear after filtration. There is limited formation of esters and other byproducts, resulting in a clean, light, and easy to drink beer. Pilsner is named after Pilsen (nowadays Plzeň), the town in the Czech Republic that became famous for the beer it produced. Most beers of the world are top-fermented beers. The principal difference between a pale ale and a lager is that all ales are top-fermented.
- *Strength (alcohol).* The alcoholic content of regular beers ranges from 3.5% to 10% and more. For heavier beers, the wort needs to have a higher sugar content. In German-speaking countries,

this is referred to as *Stammwürze*, in English as "gravity." As a measure of the sugar content, the world of beer uses the Plato scale (°P), named after the German scientist Fritz Plato (and not after the Greek philosopher). The wine world uses, for instance, "Brix," Baumé, or Oechsle. If the wort measures 12°P, this means that 12% of the wort consists of fermentable sugars, which will give about 5% alcohol. The Plato degree gives an indication of the potential alcohol. The word "potential" is used as not all sugars are necessarily fermented. Plato measures fermentable and unfermentable sugars. The so-called residual sugars obviously have an impact on the sweetness and, therefore, the body of the beer.

- *Color.* The color of beer is a function of the darkness of the malt and an indication of the role of the Maillard reaction. The color of a beer is independent from the temperature of fermentation or from sweetness; there are pale and dark lagers and ales and nonsweet and sweet dark beers. The Maillard reaction is important because it contributes to flavor richness, which implies that in general the darker beers have broader gastronomic use than lighter ones. Stouts are the darkest beers. If a very dark malt is used, the specific Maillard bitters of coffee and caramel may come into play. Stout derives its special taste from the use of roasted barley kernels (in addition to malt), which give it a typical tangy mouthfeel.
- *Bitterness.* The hops are the particular source of bitter that characterizes the flavor of most beers. The amount of bitter has its own scale. In Europe, it is called European bitterness units (EBU). There is also an international scale, international bitterness units (IBU), which is used in the United States. Both scales are only slightly different. One EBU is one milligram of isomerized α-acids, the active agent that is accountable for the bitterness.
- There are several hop cultivars, and they can be broadly divided into two groups. Bittering hops are used to impart the bitterness of beer and aroma hops deliver aroma. Provenance of hops is a second key element. The European types are more flowery and vegetal bitters, while the American and New Zealand hop species are more aromatic, reminding of citrus fruits. Lager will generally have around 5–10 IBU, imperial stout will have around 50 IBU, and India pale ale (IPA) is for the real lover of bitterness; its IBU runs up to 100.

- *Acidity.* Next to barley, beers brewed using a significant portion of wheat are the most acidic of the beer family. Examples are Weizen, white beer, or Weissbier. Wheat beers are often hard to filter due to excess proteins. Therefore, most Weizen remain unfiltered—leaving aromatic proteins and yeast in the final beer. Consequently, these beer types are often cloudy. A special type is the Belgian lambic. The difference is that, in making lambic, no yeast is added. The fermentation starts with yeasts and bacteria that are naturally present in the environment. Geuze (or *Gueuze,* written the French way) is a well-known representative of this group. It originates in the Zenne valley near Brussels. A close relative of this type is the Kriek lambic where sour cherries are added. Geuze is bottled in champagne-like bottles and is also called Brussels champagne.
- *Sweetness.* One important source of sweetness in beer has been mentioned earlier: not all sugars are fermented. The brewer has more possibilities. The most important options are adding honey, glucose syrup, or a fruit (concentrate). This leads to popular types for beer drinkers that like a beer that is easy to drink and not excessively bitter. The sweetness masks the bitters. In gastronomy, this type of beer can be served with dishes that are sweet or spicy. Although useful in that respect, beer is never as sweet as certain types of wine and, therefore, the scope is limited; many desserts are too sweet for beer.

WINE

Tea, coffee, or beer lovers may object to the seemingly excessive attention that wine mostly seems to get from gastronomes. And this book is no exception. The simple explanation is that no other gastronomic beverage is as complex and sophisticated as wine (with respect to the other beverages—although beer can get close). Furthermore, more than the other products mentioned, wine is drunk on a large scale with dishes. Therefore, it is the combination of wines and foods that demands the most attention in gastronomy. Beverages and foods do influence each other, and it is important that people enjoy what is being served. This being said, it is essential to know the flavor essentials of wine. So excuse me for dedicating more pages to wine than to the other aqueous solutions. At least those have been mentioned first.

If you do not look too closely, there is not a big difference between beer and wine. Both are alcoholic beverages that are the result of a watery solution with sugars that are being transformed into alcohol and carbonic acid by yeasts. This is about where the resemblance stops. Wine and beer are in fact rather dissimilar. We will elaborate on the differences that are important in our gastronomic context.

DEMYSTIFICATION REQUIRED

Wine is by definition made from grapes, despite the fact that there are many other fruits of which the juice can be converted by yeasts into an alcoholic beverage. To understand the flavor of wine, it helps to stay away from the traditional approach and to scrutinize common opinions and beliefs that dominate the conventional world of wine. The gastronomic view on wine first requires demystification. Especially in traditional gastronomy, it is customary to discuss wine in a somewhat mystical way. In some circles, it is a sport to recognize wines and mention as many herbs, spices, and fruits as possible when tasting a wine; descriptions are exercises in eloquence. As far as I am concerned, this is neither necessary nor wise; it creates a distance, where closeness is required. To get a grip on the flavor of a wine is difficult enough as it is and is also harder to grasp than the flavor of a dish. On the menu, the main ingredients of a dish are mostly specified and you are likely to see them on your plate, which gives an idea of the flavor to be expected—at least to some extent.

If we want to have an idea of the flavor of a wine, we primarily have to rely on the label. It basically states where the wine is from, who bottled it, and how much alcohol it contains. The region is somewhere in the world and has its specific characteristics, history, and its grape varieties that are grown there. These varieties "found their place" because they are particularly adapted to the regional circumstances. The region can be photographed and offers a wonderful reason to go there as a starting point in the understanding of the wine. But does it say much about the flavor of the wine? We will argue that the label, and therefore the region, its grape variety (or, in some cases, varieties), the alcoholic content, and even the color of a certain wine are only poor predictors of flavor. To get an impression of the flavor, you would also want to know how the wine is vinified and aged, the amount of residual sugar, or the level of acidity, but such aspects are hardly ever mentioned. In fact, it is even more complicated because qualifications such as *appellation controlée* or words such as *grand vin*, "personal reserve," and even the price of the wine can easily be deceiving; all these elements mostly promise more than reasonably can be expected. One of the cultivated mysteries of most traditionally made wines is that they are an expression of the year; consequently, they may be quite different from one year to another.

In most wine regions of the world, wine production is more or less strongly regulated by laws and regulations. Most controlled are the wines that come from what in Europe is called a protected origin, mostly referred to as an appellation. Practically every European country has its own abbreviation: AOP, AOC, DO, DOCa, DOC, DOCG, QbA, and so on. This is proudly and even compulsorily mentioned on labels. The name of the appellation stands for much more than just the origin; it often include rulings on viticultural practices, such as the grape varieties that are permitted and the minimum amount of plants that should be planted in the vineyard, training and pruning methods, yields, minimum alcohol level, winemaking methods, and so on. Winemakers in protected areas are limited in their freedom. The underlying rationale for doing so is to give assurance to consumers. They know what to expect—at least to a certain extent and if they have done their homework and learned about the specifics of the region.

The label tells the appellation: it is up to the consumer to discover what that is all about. So when you read Chianti, the mission is to know that it is in Tuscany and that the most important grape variety is the Sangiovese, but that others are permitted as well. As you go further, you will learn that not all Chiantis taste the same; there are subregions and differences in techniques and quality. There are cheap Chiantis and very expensive ones. You may even discover that some of the best wines of Tuscany—even within the borders of Chiantis—are not named as such. They do not qualify as DOCG, the highest level in Italy. They are classified as IGP or even Vino da Tavola, the lowest level, because the winemakers of those wines do not want to comply with the rules and accept the consequences. Other wine regions are no exception. The label hides many secrets and leaves much to be discovered, especially in traditional regions. The greater the difference there is within appellations, the less uniform the flavor profiles of the wines will be.

Focus on Flavor

Contrary to what is sometimes suggested, wine is not a natural product; it is a cultural product because it is made by people. A natural product would be the result of a natural process that operates without human intervention. Wine, as with all agricultural products, is the result of choices made by people along with their ideas, capabilities, and limitations within a framework that is dictated by nature. In fact, wine could be a natural product, if grapes were just left to ferment and then were bottled. This "wait and see what happens" approach is not very advisable. Something can easily go wrong in the process. The result may not always be very good, and it is certainly not predictable. Professional winemakers will not take the risk. Even the so-called natural wines that stay as close to nature as possible are an expression of the ideas of people. You may even argue that people who make these wines have stronger convictions and beliefs and may commit more, albeit different, interventions than most other professionals in the wine world.

A bottle of wine is a rich collection of chemicals; over a thousand chemicals can be distinguished. Wine consists of 95% water and alcohol; the thousand or so other chemicals make up a mere 5%. Nevertheless, that 5% makes the differences among wines. The focus on the intrinsic flavor qualities helps to form a practical understanding. Mouthfeel and flavor richness provide useful answers to practical questions about wines, such as when and with what to drink them. What we need to do is to analyze where and how the flavor factors are affected, or rather to interpret whatever is in the glass. We should do the same as we did in Chapter 4: follow the path of wine and see where decisions are made that influence its ultimate flavor. There are many moments in the process of winemaking where such choices are made. In fact, it all starts before wine can be made: in the vineyard where the grapes are grown. In the next sections, we will look at the respective influences of:

- Grapes
  - Terroir, climate, and grape variety
  - Agricultural choices and practices
  - Harvest

- Vinification
  - The role of the skins
  - The role of yeasts
  - The role of temperature
  - The role of the lees
  - The role of malolactic fermentation
  - The role of wood

- After vinification
  - Blending
  - Further aging
  - Marketing (bottle, label, price, distribution)
  - Serving

THE EFFECTS OF GRAPES AND GRAPE GROWING ON THE FLAVOR OF WINE

TERROIR, CLIMATE, AND GRAPE VARIETY

Grape varieties are grown where they are for a reason. Mostly they are there because other plants or other uses of the land are not possible or not as attractive. In traditional wine regions, the choice of certain grape varieties has gradually settled down. By trial and error, clonal selection, and breeding, varieties have been selected that perform best in a specific region or plot. The French use the term *terroir* for the all-encompassing concept of regional climate and microclimate, composition of the soil, availability of water, fog, and wind, the difference in day and night temperature, the hours of sunshine and luminosity, and so on. The best grapes are harvested from plants that are perfectly suited to the terroir. Therefore the terroir is often mentioned as a major contributor to flavor. Some even say that the best wines are an expression of the terroir. Although, as we will see, this assumption is somewhat dubious, it is certainly true that you need good grapes to make good wine. Unfortunately, it is also possible to make a terrible wine from the best grapes. In cooking, this is no different.

Finding and selecting the right grapes and clonal selections is an ongoing process. Some countries are just entering the world of wine. Wine grape growing is increasing in cool and wet regions such as in the Netherlands, Denmark, and Great Britain. The supposed climate change has little or nothing to do with the development of wine growing in areas that were previously not very suitable. The reason for this expansion is the existence of successful new hybrids that are available now, such as the Regent, Johanniter, and Solaris. The challenge for these countries is to find out what the real qualities of the new plants are and to select the best variety for the soil. It will take years before the best matches are found. You can only harvest once a year, and it is only after a few years that the grapes of new vines can be used. And when wine is made from them, it takes years to see how they develop and what techniques can best be used. Fortunately, experience and knowledge become more and more public, which means that winemakers can learn from each other and from publications.

It is not just in the new countries that this process is going on. Even in countries that are better suited for growing grapes and those with more experience, there is a continuous process of selecting good grapes, replacing others, finding new plots, and so on. In some regions, the possibilities are severely limited, but in countries that are less established such as South Africa or Argentina, there is generally a strong focus on experimenting with varieties in order to find the best matches of terroir and variety. In South Africa, mountains and ocean breezes are major influences. In Argentina, growing at a high altitude becomes more and more interesting. Certain grapes, such as Sauvignon, Chardonnay, and Pinot noir, develop or keep their aromas better. In fact, there are many variables involved, which means that it takes time to select the "right" variety or varieties.

The relationship between grape varieties and climate is often condensed into three categories: there are cool, mid, and warm climate varieties. This is based on the average growing season temperature:

- Cool climate: 16.5°C or below
- Mid (or moderate) climate: between 16.5°C and 18.5°C
- Warm climate: between 18.5°C and 21°C

It is generally asserted that the cool climate types are the most aromatic. Certain vinification techniques are used in order to retain these specific aroma characteristics. The temperatures are kept low, and oxidation is prevented as much as possible. These wines are mostly drunk when they are still young. As they age, they tend to lose their specific character. Malolactic fermentation and oak aging also are not used as much for the same reason. However different, mid and warm climate varieties can be aromatic as well. The Muscat or moscato is probably the best example. Winemaking choices will be in line with what is best for the cool climate types. Chardonnay (mid climate) may be the best example of a variety that does not excel in specific aromas. The flavor of Chardonnay wine is therefore dominated by choices that are made in vinification. Such flavors are called "secondary flavors." Primary flavors are the ones that relate specifically to grape character. Tertiary flavors come about by aging.

AGRICULTURAL CHOICES AND PRACTICES

After the right plants for the situation have been selected, the role of the farmer is crucial in growing grapes. He needs to work the land, prune the vines, provide protection for the plants, and create the conditions that make the grapes grow well. He needs to do all this in order to be able to harvest. Every farmer needs to make a choice about how he wants to work. He can make his life easier by relying on chemicals as fertilizers, insecticides, and fungicides. There is a strong tendency to make organic choices or to at least limit the amount of chemicals as much as possible. Among the best grape growers there seems to be a tendency to adhere to the biodynamic principles that go way beyond organic farming. Whatever choices are made, they have influence on the grapes that are ultimately harvested as well as the microorganisms (such as yeasts) that grow on them.

Besides keeping plants healthy, the mission of grape growing—or any other fruit growing for that matter—is to get the grapes to the desired level of ripeness. Clearly, this has everything to do with accumulating sugars. These are formed by photosynthesis, which means that the green leaves of the plant are extremely important. They are the plant's motor via chlorophyll. There is an important relation between sunlight and heat and the amount of leaves. In cooler regions, more leaves are needed to produce good grapes than are needed in warmer and lighter regions. Deciding on the amount of leaves on the plant is called canopy management. Looking at the amount of leaves in a photograph, it is often not very difficult to guess if it was taken in a cooler or warmer wine region.

Another most important aspect of the farmer's work—given that the mission is to produce the best grapes—is to limit the yield. This is the paradox of grape growing: in most cases, the goal is not to produce as many grapes as possible but to find the optimum balance between quantity and quality. Yields can be reduced in several ways, and one of them is fundamental: how many plants are planted per hectare/acre. The more plants there are, the more rootstocks are in the ground, and the more nourishment goes to the grapes (provided that there are fewer bunches of grapes per plant). In classical regions, such as Bordeaux and Burgundy, the density is up to and even more than 10,000 plants per hectare—on every square meter there is a plant. In practice, this may mean that there is only one bunch of grapes per plant. Yield per hectare is often used in the world of wine and usually regulated by regional wine laws. Understandably, this yield adds an extra dimension if there is also a high density of plants. In some regions, such as Rioja in Spain, there is plainly not enough water to have a high density; therefore, there are fewer plants per hectare in these parts.

The availability of water is an interesting subject. It is often said that the vine needs to suffer to produce the best grapes. This is true to a certain extent as far as the rootstock is concerned; it must be encouraged to grow deep into the soil and rocks (especially when the soil is rich in minerals). Yet, it is certainly not true for the leaves: they need water—although not too much. However, in many European regions with an appellation, irrigation is not permitted. The vine has to survive by itself, and in most years it will have no problem doing so. In some years, however, there is drought, and this is a severe problem. Water stress blocks the growth, leaves turn yellow, photosynthesis stops, and no sugars are formed in the fruit. This means that in extremely warm and dry years, the quality of the wine is rarely good. In regions that are less restricted, farmers often use drip irrigation. Modern computerized systems monitor exactly how much water is needed and supply that to the plant.

The quantity of grape bunches on the plant is also regulated by pruning. Bunches are formed on new shoots. Ultimately at the end of the winter, when the plant is still dormant, the vines are cut back and a specific number of buds are left. The fruit is formed on the shoots that start to grow in spring. After the flowering and pollination, the bunches are formed and the farmer gets a first impression of the potential quantity of the harvest. If everything develops well (no hail, frost, or outbreak of diseases), he may decide to cut away bunches before they enter the last phase of ripening. This is often called the green harvest (*vendange verte*) and is done to obtain better quality grapes. All the nourishment that the roots provide goes to the grapes that are left.

Specific conditions in regions may lead to a specific character in grapes. Temperature, humidity, and wind are important. Humidity is a danger because it may lead to all kinds of negative developments. Principal among these are fungal infections of the plant such as mildew and gray rot (*Botrytis*), which have a negative impact on the grapes. If not treated, such conditions can destroy a harvest. In one specific case, however, rot can actually be a blessing in disguise. Natural circumstances such as sunny afternoons or warm winds may prevent rot from occurring or—even better—may promote the development of a special type of fungus, the noble rot (*Botrytis cinerea*), *pourriture noble*, or *Edelfaüle*. The skins become permeated, lose some moisture, and pectins are formed. The wine of such grapes is sweet and can often be recognized by touches of honey.

HARVEST

When the grapes are ripe, one of the most far-reaching decisions needs to be made: when to pick. This determines the flavor of the wine that can be produced. A lot happens in the last phase of ripening. Most apparent are the rise of the sugar level and the decrease of acidity. Both are necessary for making good wine. Sugar is important for forming alcohol. About 17 grams of sugar produce one gram of alcohol. Therefore, it is important to know the sugar content of the must, or must weight. German wine-oriented countries will generally use Oechsle degree as an indicator; the winemakers that are educated in France will mostly use the scale of Baumé, whereas the Americans use Brix. However different, every scale does the same thing: it informs the winemaker of the potential alcohol. Table 5.2 shows the differences.

The acidity of the grape is important as well. The levels drop during ripening and their composition may change from malic to lactic acid. In warm regions or years, grapes may even lose all malic acids. This is the main development; there are other transformations in acids as well. It has already been mentioned that different acids have different flavors in type and in mouthfeel. Consequently, the winemaker not only follows the development of the sugars when choosing the date to start harvesting, (s)he also analyzes the development of the acidity.

There is a third important component to watch. And this seems to be considered the most important these days: the phenolic ripeness and the maturity of the skin. This is when the aromas are formed and tannins become ripe. An important indicator is the color of the seeds. In ripe fruit, seeds are black. (By the way: do you often see black pips in apples that you buy in the supermarket?) In vinification, sugar and acidity can be corrected in some way, but the aromas cannot. This is illustrated in Table 5.3 with the evolution of aromas in Cabernet Sauvignon grapes. In the early stages, the green and unripe flavors predominate; likewise, the tannins of the skin will be unripe and green. The wine of such grapes will not have the mouthfeel or the aromas of a wine that is made from ripe or overripe grapes. (The precise nature of this aspect of ripeness will be discussed in the section "The Role of the Skins" in this chapter.)

TABLE 5.2
Measure of Degree of Sugar and Potential Grams of Alcohol

Oechsle	Baumé	Brix	Potential Alcohol (gr)
65	8.8	15.8	8.1
70	9.4	17.0	8.8
75	10.1	18.1	9.4
80	10.7	19.3	10.0
85	11.3	20.4	10.6
90	11.9	21.5	11.3
95	12.5	22.5	11.9
100	13.1	23.7	12.5
105	13.7	24.8	13.1
110	14.3	25.8	13.8
115	14.9	26.9	14.4
120	15.5	28.0	15.0

Source: Halliday, J., Johnson, H., *The Art and Science of Wine*, Mitchell Beazley, London, 1992. With permission.

Clearly the style of wine that the winemaker wants to make influences the choice of when to harvest. To produce quality wine, the winemaker really needs aromas. These are formed in the last phase of ripening, after sugars have reached their peak and acidity is at its low. This choice explains the character of many modern wines: high

TABLE 5.3
Development of Aromas in Cabernet Sauvignon Grapes

Ripeness of Cabernet Sauvignon Grapes					
Vegetation	Herbaceous-ness	Unripe Fruit	Red Fruit	Black Fruit	Jam
plant matter	straw, herb, vegetal	green apple, citrus rind	cherry, strawberry, raspberry, cranberry	plum, blackberry, black cherry	prune, date, raisin, tobacco

Source: Bisson, L., *Practical Winery and Vineyard Journal*, 32–43, July/August 2001. With permission.

in alcohol and they sometimes even seem to be on the sweet side. This is partly explained by the relatively low acidity. Nonsparkling mineral water with a low mineral content may seem sweet for the same reason.

Another aspect of harvesting is whether it is done by machine or by hand. Hand picking is often recommended as the best way to gather grapes, but the reality is that in most cases harvesting machines are used. These have become more and more sophisticated. There are even systems these days that are guided by satellites that monitor the ripeness of certain plots. The riper grapes are automatically put in different baskets. In specific cases, hand picking must be done. In some regions, such as in Champagne, mechanical harvesting is simply forbidden. The local regulations require that whole bunches are put in the press and determine how many liters can be extracted. In Champagne, the allowed must maximum is 102 liters from 160 kg of grapes.

In other regions, such as in the Douro or the Mosel or Rheingau, machine harvesting is plainly impossible. In Germany and Austria, there is another reason why machine harvesters are not much used: their wine laws have separate categories (*Prädikate*) for grapes that are harvested at a specific degree of ripeness. *Spätlese* stands for late harvest; in the next levels, the word *Auslese* is used. This refers to selecting grapes. *Beerenauslese* and *Trockenbeerenauslese* are the ripest selections of grapes. In practice, this means that in these countries or regions, for some of these quality levels, whole bunches are not harvested but specific grapes from bunches. Very late harvested grapes are often touched by noble rot, or they have dried out and are harvested more like raisins. The French word for this dried out condition is *passerillage*. In dry and windy regions, this can be promoted in the plant by cutting away leaves to expose the grapes to the sun. This process of raisining (drying of grapes) can also been done after harvest. Bunches of grapes are left to dry. The wine that is made from such grapes is called straw wine, *vin de paille*, or *recioto*. With or without *Botrytis*, the grapes have a higher sugar content than yeasts can handle. As a consequence, such wines are sweet.

THE EFFECTS OF VINIFICATION
ON THE FLAVOR OF WINE

THE ROLE OF THE SKINS

Phenolic ripeness has already been introduced as an important contributor to flavor. The word *phenolic* refers to a large group of chemical compounds that affect the color, flavor richness, and mouthfeel of a wine. This group broadly falls apart into flavonoids and nonflavonoids. The most important flavonoids include the anthocyanins and the tannins. The anthocyanins give the color; the tannins are astringent compounds that affect mouthfeel. The principal nonflavonoids are phenolic acid and the stilbenoids. Phenolic acids are particularly active in forming aromas in vinification. A specific stilbenoid is resveratrol—the compound that is associated with the positive health effects of red wine.

Certain phenolics, especially precursors of aromas, must be extracted before the vinification actually starts. The crushed, not yet pressed grapes are left to stand in a cool place for some time. This is called skin contact, or in French, *macération prefermentaire* or *macération pelliculaire*. Cooling is necessary to prevent an undesired spontaneous start of fermentation. These days it is common to add enzymes in this phase—pectinases, hemicellulases, glucanases, and glycosidases. They break specific molecules into smaller components: for example, glucanases break down glucans such as starch, cellulose, or laminarines into glucose molecules. Each type of glucan has an enzyme of its own. In the process, molecules become free that would otherwise play no role in flavor richness or in mouthfeel. The skins become soft, which also speeds up the vinification time.

In vinification, the skins have a major role: they contain the pigments that make the color. The color of a wine is a function of the time that the skins are left in contact with the must and of the time of extraction. Consequently, white wine can also be made from red grapes, provided that the grapes are immediately pressed after harvest so that the skins get no time to color the must. Rosé can be made by leaving the skins with the juice for some time; for red wines, the extraction needs to be longer. The colors are extracted in a few days, but there are more phenolic compounds. Grape varieties differ greatly in phenolics. The anthocyanins give the color. Some red varieties have more pigments than others. The Pinot noir is known for a low level of anthocyanins. The wine of this grape will generally be light in color. On the other hand, the same grape is very rich in all kinds of other phenolic compounds.

A longer extraction will also increase the level of tannins and certain aromatic precursors. Therefore, the winemaker wants to look for a balance. It is similar to the steeping time of tea. The longer the extraction, the more phenolic compounds will become part of the beverage. The wine will be pressed as soon as enough colors and tannins are formed; mostly the pressing is done when the wine has not yet completely finished its fermentation.

THE ROLE OF YEAST

The skins not only have a quality of their own; the surface of the skins contains a large variety of molds, bacteria, and yeasts. These microorganisms are there by the millions and not only on the skins; they are in the air and in the cellar as well. Yeasts deserve our special attention because they are the one and only true winemakers. Yeasts form a group of about 160 species. Those that belong to the genus *Saccharomyces* are particularly important in gastronomy. They are the ones that give us our bread and our alcoholic beverages.

As soon as the grapes are picked, the yeasts would love to start converting sugars as soon as they can get to them. This is the natural way of forming ethanol and CO_2. During the fermentation, besides alcohol, many other new aromas are formed. The winemaker is likely to want to manage this process; he wants to make a certain type of wine and to be in control. To that extent he uses SO_2; it stops all biochemical action. Furthermore, he will keep the temperature well under 10°C; yeasts are not very active at low temperatures. For the same reason, especially in warm regions, harvesting is mostly done at night. To make white wines, the pressing is done as quickly after harvest as possible. When distances are an issue, mobile pressing units can be used because grape juice is easier to control and to transport than grapes.

The commercial winemaker wants to have more predictable control of fermentation and quality and to create a reproducible and stable product. Therefore, the common practice is to add a mixture that helps the must to start fermenting and to add commercial strains of *Saccharomyces cerevisiae*. The winemaker makes a deliberate choice because the role of yeasts is not limited to the metabolism of sugar; the fermentation has a big influence on the overall flavor of the wine—on mouthfeel and on flavor richness. The winemaker generally buys his yeast from one of a few companies that produce all the yeasts for the world. On their websites,[5] he reads about the effects of every yeast species or strain. Some give a certain aroma, others enhance mouthfeel, and some are more suited for fermentation at lower temperatures, others at higher. And every yeast strain has a maximum of alcohol that it can handle. Some reach levels of more than 16% ethanol. This explains—at least to some extent—how wines can have so much more alcohol these days compared to, say, those of the 1970s.

This book will not go into depth about how yeasts (inter)act and address all of their characteristics and functions. However, it must be said that in contrast to what is stated in many regular wine books, is not just the *Saccharomyces cerevisiae* that do the work. In fact, fermentation is a complex process where many other families of yeasts and other sub-stances are active. Yeast species, including non-*Saccharomyces* organisms, deliver a vast array of flavor-impacting substances such as organic acids, glycerol, higher alcohols, esters, aldehydes, ketones, amines, and sulphur volatiles. All of these have a certain impact on the character of the wine.[6,7]

The logical consequence of all of this is that the selection of yeasts is a complex matter and a part of science that is in full development. The wine world needs to discover the complexity of the fermentation process. If just one or a few strains from the genus *Saccharomyces* are used, the winemaker risks ending up with a one-dimensional wine. Or a wine that has little or no regional expression and tastes like a wine that is produced in a different part of the world. There appear to be large regional differences among yeast families. This adds a new and fascinating dimension to the concept of terroir. It makes you wonder about the future role of AOP ruling agencies; they may want to consider including the use of the local yeasts in their regulations.[8,9]

These findings promote research on the fermentation process and on agricultural practices that keep wild yeasts alive or that even stimulate the growth of desired microorganisms. When the original, indigenous yeasts of the grape and environment were used, nature took care that all microorganisms were present to make the wine. Knowing all this, it can no longer be a surprise that, all over the world, winemakers are reconsidering the use of wild, indigenous yeasts. A region to mention in this respect is the Mosel in Germany. It is interesting to note that the most avant-garde of winemakers has realized huge progress in the quality of their wines by going back to nature. Pioneers to mention are Markus Molitor in Zeltingen and Reinhard Löwenstein in Winningen. Both work together with scientific research institutes to resolve the mysteries.

THE ROLE OF TEMPERATURE

Temperature control is a standard feature of modern winemaking. The days when water was sprayed over vats if temperatures got too high are long gone. Fermentation produces heat, and the rise of temperature has always been a matter of concern. Temperature influences the formation of certain volatile aromatic compounds. To preserve aromas and certainly the primary fruit flavors, the temperatures should not get too high. In one study, for instance, primarily fresh and fruity aromas were produced by keeping the vinification temperature at 15°C, while higher concentrations of flowery-related aroma compounds were found at 28°C.[10]

Yeasts operate within certain temperature brackets. The regular wine books will state that they can only operate above 16°C and will die if the temperature rises above 35°C. Now that we know more about yeasts and the other microorganisms that are active during fermentation, this needs reviewing as well. Yeasts can be active at lower temperatures. On the high side, the number is quite correct; to prevent the loss of primary aromas, there is a general tendency to keep the temperatures below 30°C in any case. White wines are mostly vinified below 20°C. Furthermore, yeast strains have their own temperature at which they perform best,[11] as we have also seen when discussing the brewing of beer.

Temperature not only affects the aromatic structure of wine. The so-called secondary products of fermentation (glycerol, acetic and succinic acids, and acetaldehyde) increase as fermentation temperatures rise.[12] Glycerol is a substance that influences mouthfeel; succinic acid is used in the food industry as a sweetener. Together with a molecule of ethanol, the latter forms mono-ethyl-succinate, which gives a mild fruity flavor. Acetaldehyde gives a green apple aroma.[13] When it is oxidized, it turns into the aroma that is called maderization, after the wine of Madeira that traditionally oxidizes in wooden casks in a warm environment (not the cellar, but the attic). These are just a few of many examples illustrating that the flavor of wine is the result of a complex process of chemical reactions. Grapes are just a part of the story.

THE ROLE OF THE LEES

After the work of yeasts is done, they become inactive and settle down on the bottom of the wine container. These "dead yeasts" constitute a large part of what is called the lees. The word "dead" is deceiving. The lees are not inert. They can still have an important function, especially in flavor. In Muscadet, this has even led to adding *sur lie* on the label to recognize that this specific class of Muscadet is generally richer and fuller in flavor than the regular type. Where does this richness come from?

There are two obvious elements of wine that contribute to mouthfeel: alcohol and residual sugar. Does this imply that all coating wines are either rich in alcohol or have some level or form of residual sugar? Clearly not. For instance: a category of wines that is from Chardonnay or Pinot blanc and also some sparking, rosé, and red wines may be classified as coating, although they are completely dry and have a moderate level of alcohol. There must be a third, less obvious source of coating mouthfeel. In the world of food, we know about the importance of umami; it brings flavor richness in a coating form. As we discussed earlier, umami is the name of the flavor of a specific amino acid: glutamic acid and some 5′-nucleotides, IMP, adenylic acid (AMP), and GMP[14] (see also the sections "Umami" in Chapter 3 and "Factor 4: Presence of Umami" in Chapter 4).

Can such components be found in wine as well? If so, to start with, it must be established that glutamic acid is indeed one of the building stones of the proteins of grapes; that this amino acid is still there after fermentation; and finally, that it can be set free by enzymes or otherwise. The presence of GMP could also be studied because this nucleotide can be found in plant-based foods. And finally, can proteins and peptides make a significant contribution to wine flavor, knowing that they are only minor elements of a wine? This kind of deductive thinking led to the discovery that umami is indeed present in wine and that it is the component we are looking for.[15]

The Pinot family of grapes (Chardonnay, Pinot blanc, Pinot gris, Pinot noir, Pinot meunier) is particularly rich in glutamic acid, and it does not get lost in fermentation. To liberate it, autolysis is required; and this has everything to with the lees or, more specifically, with stirring them or otherwise keeping them in suspension. Autolysis plays an important role in winemaking. It activates enzymes, which leads to a degradation of cell proteins. This results in a variety of cellular elements that have an influence on the flavor of the wine such as amino acids, peptides, fatty acids, and nucleotides, and the release of soluble mannoproteins from the cell walls. Wines with these elements have a more coating mouthfeel and more flavor richness, just like the umami effect in food.[16,17]

Umami-rich foods are popular, and the same tendency is seen in wine preferences. An intensified focus on the role of the lees in winemaking is therefore to be expected. The word *bâtonnage* in French (or "stirring the lees" in English) is heard more and more often and not only with regard to white wines. Aging on lees also appears to be a good technique for improving red wines. It enhances some fruitiness and makes them rounder and "fleshier." At the same time, the astringent and bitter side of tannins is less apparent. However, there is one thing to keep in mind: the autolytic process is very slow, especially at cellar temperatures. Research today is focused on how the process can be speeded up.

To conclude, next to white and red wines, there is another class of wines that must be mentioned with regard to autolysis: sparkling wines that are made following the rules of the traditional method that originates in Champagne. The CO_2 is formed in the bottle by a second fermentation, which implies that such wines all age *sur lie* and the longer the better. The AOP rules of Champagne require that the wine should age at least 12 months on the lees. Quality producers cellar their wines longer; and for the best wines of the region (those that are made exclusively from one harvest year, the *millésimés*), this minimum is three years. The reason for this is because it is only after two years of aging that the concentration of 5'-GMP slowly increases. In practice, the most prestigious wines are kept much longer on the lees. Together with the presence of free glutamic acid, this explains the flavor richness of these wines, especially when you compare them to similar wines that have not been in contact with the lees as long.[18]

THE ROLE OF MALOLACTIC FERMENTATION

Whereas research on the role of the yeast and the lees is in full gear, the impact of malolactic fermentation (MLF) on flavor has been known and studied for a long time. MLF does not involve yeasts but is in fact a bacterial process that converts malic acid into lactic acid. These two acids are perceived quite differently. Lactic acid is softer than malic acid.

MLF can therefore be considered a natural method of deacidification. One carboxyl group of malic acid is removed, which yields lactic acid. Most red wines undergo MLF and some of the white wines, mainly the ones that are considered to be too high in acidity.[19]

In the past, MLF happened more or less by accident in the spring when temperatures in the cellar rose, causing the bacteria to start developing. These days the process is under control and starts by adding specific bacteria (e.g., *Oenococcus oeni*, several species of *Lactobacillus*) to the wine. Again there is every reason for control as the influence of MLF on flavor does not stay restricted to the conversion of acids; aromas are influenced as well. The ripe flavor tones, especially, are enhanced. After MLF, Chardonnay wines develop notes of hazelnut, fresh bread, and dried fruit aromas. In Pinot noir wines, the animal and vegetable notes increase. Wines of the same grapes without MLF kept specific aromas such as apple and grapefruit-orange in Chardonnay and strawberry-raspberry in Pinot noir.

In general, wines that have undergone MLF have higher concentrations of diacetyl, which produces a characteristic buttery flavor. In general, flavor characteristics associated with MLF are described as floral, buttery, nutty, yeasty, oaky, sweaty, spicy, roasted, toasty, vanilla, smoky, earthy, bitter, ropy, and honey. Clearly, these characteristics do not fit all types of wines. Winemakers use this knowledge to maximize or minimize a particular flavor attribute. To mention an example: the buttery flavor of diacetyl does not match well with Riesling or Sauvignon. Therefore, MLF is to be avoided with these grape varieties.[20]

THE ROLE OF WOOD

Wherever in the world they are made, the best white and red wines and many wannabes spend some time in a wooden barrel of—mostly—French oak. One of the big misunderstandings of wine is that wooden barrels are supposedly used to give wood flavor to wine, and especially notes of vanilla, toast, and sometimes even coffee. The same aromas have been mentioned before and, therefore, should not and cannot specifically be attributed to oak barrels. Furthermore, the best wines of the world want to excel in a specific and often regional character. It is not likely that winemakers are looking for an extrinsic element that subsequently is going to dominate the subtle characteristics that they have just obtained by growing the best possible grapes and controlling the fermentation as much as they can. Yet they practically always do use wood; therefore, the aging in wood—and even vinification in some cases—must apparently have a desired influence on the flavor of wines other than giving woody character. There must be an explanation why.

A simple answer cannot be expected. In general, oak barrels contribute to the flavor richness and to mouthfeel through the phenolic compounds of the wine, particularly tannins and lactones. Lactones have aromatic potential. Additionally, wood aging influences the intensity and stability of the color of the wine. There are many elements involved. The flavor of the wine is influenced by:

- *The type of oak that is used.* There are three species of oak that are particularly suited for barrel making: *Quercus sessilis*, *Quercus pedunculata*, and *Quercus alba*. The first two are the ones that are found in France—especially in the center of the country in the forests of Tronçais, Allier, Limousin, Burgundy, and Vosges. There are other regions in Europe where useful oak can be harvested, but the French wood is still the most precious. *Quercus alba* is the white oak that you see often referred to as American oak or Tennessee white oak. The difference is mainly in the grain of the wood. The finer the grain, the more subtle the influence of the barrel. *Quercus pedunculata* oak produces much tannin and few aromas; these barrels are mainly used for spirits. White oak does not produce much tannin, but it is very aromatic and brings a strong vanilla character. *Quercus sessilis* is the most balanced.
- *The size of the barrel.* When the word *barrel* is used, it is generally a cask of 225 liters, which represents a volume of 300 bottles. In Burgundy, the normal size of a cask is 300 liters, and there are many other sizes, up to thousands of liters, although these are not called barrels anymore but casks (*foudres* in France, or *großen Holzfass* in Germany). Clearly the influence of the wood decreases as the size increases. There is a renewed interest in using wood, especially among winemakers that want to produce high quality

wine. The phenolic development requires oxygen, which promotes the use of wood instead of inert materials such as stainless steel.

- *The toasting.* Making barrels is an ancient craft that involves a careful selection of dried and matured oak and also involves the use of fire. Heat is used to bend the staves. Much needs to be done by hand. The time that the barrel is over the fire has a strong influence on the formation of aromas. A light toast implies that the barrel has been heated about five minutes, just enough to make the barrel. It gives a real oaky, wooden aroma. Medium toasted barrels have been heated for about 10 minutes at 180°C–200°C. For medium plus, this time moves up to 15 minutes. As the time and temperature increase, the aromas get more complex, less woody, and more vanilla, roasted bread, and spices. Heavily toasted barrels are over fire for about 20 minutes, and the temperature rises to about 200°C. Aromas change into smoky, caramel, and coffee.
- *The time that the wine is in the barrel.* Barrel times differ from wine to wine and from year to year. Winemakers scrutinize the development of their wines in the barrel and taste them regularly to decide how long wines are to be kept in wood. Rich, red Bordeaux type wines are generally kept in the barrel for 18 months.
- *The number of times that the barrel is used.* New oak barrels have more impact on flavor than used barrels. Every year that a barrel is used, the influence decreases. Therefore, in many cases, the winemaker is likely to use a mixture of new, first-, and second- (harvest) year barrels. The proportion can change from year to year and is dependent upon the primary quality of the wine. The better the basis, the more new wood and the longer the time that the wine may stay in the barrel.

These are all instruments that the winemaker can play with. The use of wood is or should be a function of the wine (whether it is red or white) and the grape variety. Aging in wood alters the phenolics in the wine under the influence of oxygen. Tannins and anthocyanins polymerize (form longer strings) and the flavor becomes less astringent. The amount of tannins and aromas that are formed and the speed of the process are largely dependent on the factors mentioned here. The compounds related to oak toasting (guaiacol, 4-methylguaiacol, furfural aldehydes, and vanillin) are all reported to be extracted faster during the first days of oak maturation—except for vanillin, which required at least three months to accumulate in the wine. The volatile phenols, 4-ethylphenol and 4-ethylguaiacol, are formed in large quantities after the first 90 days of oak maturation. Wines matured in large oak barrels of 1,000 liters resulted in the lowest levels of volatile compound accumulation.[21]

Oak barrels are expensive, and cheaper alternatives are used extensively, especially wood chips and staves that are added to the wine for some time. Superficially, the use of these materials gives adequate results; wood flavors are added, but there are no complex interactions between the wine and the barrel. Questions have been raised if whether the use of these substitutes is in fact not simple ways of aromatizing wines. When this is allowed, why not allow adding other aromas as well—a concentrate or powder of black currant, strawberry, raspberry, vanilla, or tobacco, perhaps?

AFTER VINIFICATION

BLENDING

It is quite common that at first not one but several base wines are made. They come from different grape varieties or plots. There may also be differences in vinification and/or aging. Some of these wines will have MLF or pass some time in a barrel, while others will not. Every batch of wine represents a *cuvée*. These batches are mostly kept apart until the end when the final blends are made, which implies that the winemaker still has some tools to manipulate the flavor of the final wine, even after it has been made. These possibilities fall under the general topic of blending, mixing several *cuvées* together.

There are several possibilities. The first is blending different wines of different grape varieties. Great wines are often blends. The AOP of Bordeaux allows several grape varieties; the most important are

Merlot, Cabernet Sauvignon, Cabernet franc, Malbec, and Petit Verdot. In most cases, this means that the Merlot is blended with one or more of the other ones. Every year the winemaker is able to make the best blend possible. If the Cabernet Sauvignon does not reach full maturation in a certain year, other grapes can be used to compensate. Furthermore, every variety brings its own specific character to the wine. The Merlot is assumed to be somewhat more coating than the Cabernet types. The Cabernets are more tannic and bring aromas of black currants and spices. A good blend of two or more varieties will add to the complexity of the wine. A similar situation can be found in the southern part of the Rhône Valley, where the Grenache replaces the Merlot in the equation. It can, for instance, be blended with the Syrah, Mourvèdre, Carignan, or the Cinsault; and each one will add its specific character. Some AOPs even allow a percentage of white wine to be added to the red wine. A famous example of this practice is Côte Rôtie in the northern part of the Rhône Valley, where up to 20% of the white Viognier may be added to the red wine that is made from Syrah grapes.

Blending may well go further than grape varieties. If a certain wine domain has plots with differences in terroirs, the wines of these plots may be kept separately. Château Margaux, one of the famous first growths of Bordeaux, has 26 distinct *cuvées*. The final blend is a yearly challenge. Wines that are not used in the first wine are used for the "second wine," which is called *Pavillon rouge* or even a "third wine." This system allows rigorous selection in an effort to make the best possible first wine. The differences in the plots can be due to differences in location, soil composition, drainage, exposition, and so forth. Newly planted plots can also be kept outside the definitive blend. The grapes of young vines may lack complexity in their early years.

A third type of *cuvée* may be wines that are made slightly differently. A white wine may be a blend of a wine that has undergone malolactic fermentation and a *cuvée* that has not. As we have seen, such a wine will be more acidic and different in aromas. The same can be done with wood aging. Making a blend of wines with or without MLF or barrel aging is a way of adding flavors to the final wine.

FURTHER AGING

The finished wine is bottled and put on the market. Throughout the world, most wine is consumed within 12 to 18 months after it is produced. The vast majority of wines are drunk shortly after having been purchased. This has several implications. From the perspective of the winemaker, it is important to bring wines to the market that are ready to drink. More and more is known about the factors that influence the flavor of the wine, so the wine business is getting better and better at bringing wines on the market that are pleasant to drink. Some regions have wine laws that regulate cellaring time for that reason. Champagne has already been mentioned; Spain and especially Rioja have special labels for wines that have had a certain amount of aging, in barrel and in bottle. Highest in rank are the Gran Reservas. Wines with, respectively, the labels "Reserva" or "Crianza" have not been aged as long. Traditional regions such as Bordeaux and Burgundy never had such regulations. Instead they instructed the buyers to cellar their wines for a long time and drink them years after bottling—especially because the tannins needed time to soften up.

Although much has changed—even in the traditional regions—most people still assume that the longer you keep a wine, the better it will get. This is only true in some cases; most should be drunk within a few years after harvest. Classic red wines made from Cabernet Sauvignon, Syrah, or Nebbiolo are examples of reds that generally will age well. The unexpected exceptions are top quality whites, especially if they are made from the Chardonnay, Riesling, or Chenin blanc. These whites may need time to show their real potential and are often drunk too young. In any case, it is important to store wines well, at least somewhere cool and not too dry. If aged too long, all wine will pass its peak and eventually turn into vinegar—although very old wines, way past their peak, can still be enjoyed out of curiosity.

Oxidation is a wine's enemy and a wine's friend. When oxygen reacts with alcohol, it creates acetic acid, or vinegar. Also, in aging, the color of wines gradually changes. Red wines turn from purplish into orange-brownish. In contrast, white wines get darker. Their color develops from lemon green and straw into golden, honey colors. Very old whites, and especially sweet ones, may turn amber and orange as they oxidize and age. It is similar to apples turning brown after they have been peeled and sliced. Just as a little lemon juice keeps the cut apple from browning, wines with higher acidities tend to turn brown less rapidly. In general, wines with a relatively high acidity are more suited for aging. The changes in colors of wines are—just as in a barrel—caused by a polymerization of tannins and polyphenolics. They bind together in long chains and eventually form larger molecules that fall to the bottom of the bottle as sediment. The logical consequence is that such aged wines lose astringency, get less dry, and are perceived as softer. In contrast, sweet wines that age gradually will be perceived as less sweet, even though nothing changes in terms of the actual grams of sugar.

To develop positively, wine needs oxygen. For that reason, corks are a good way to close the bottle. Wine "breathes through the cork," and this helps the wine to mature. However, knowing that most wines are not meant to be kept or to age, it is better to use one of many modern inert materials for all these wines. The best corks can then be reserved for the best wines of the world that truly improve in the bottle.

MARKETING (BOTTLE, LABEL, PRICE, DISTRIBUTION)

The title of this section could also have been "extrinsic factors." These fall under the general heading of marketing, and it would be silly to assume that making a good wine is all that it takes. The choice of a bottle form and color, label, price, and even distribution are factors that influence the perception of the wine. To be successful in the consumer world, a winemaker should understand the intrinsic as well as the extrinsic factors. The importance of extrinsic factors has been stressed before. In boutique wine stores or in restaurants, useful information on the wine can be given by a trained professional. But the bulk of wine is sold in supermarkets, and by definition there is little or no help in selecting wine; the "outside" has to do the job of convincing consumers to buy the wine. For that reason, labels and bottles are important in that they convey a message about what is in the bottle. Some regions have developed their own engraved bottles that area winemakers are obliged to use. A good example is Châteauneuf-du-Pape.

The world wine market has also become significantly more competitive. France, Italy, and Spain combined produce slightly more than half of all of the world's wine, but in the past 30 years, their own per capita consumption has fallen by 40%–50%. At the same time, competition emerged from the so-called "new world" nations of South Africa, Chili, Argentina, Australia, New Zealand, and the United States. This globalization has gone hand in hand with the availability of information through the Internet. The prosperous wine consumer of today is more knowledgeable and has a more sophisticated understanding of the product and its value. Success as a wine producer in the twenty-first century requires a thorough appreciation of human behavior and product choice. Prosperous consumers chose quality rather than quantity in consumption. It is quite likely that new wine consumers or less knowledgeable ones are more susceptible to extrinsic factors, such as label, brand, and price.[22]

It is good to know about consumer preferences. It is reported that females generally drink less wine than males, prefer white over red, and medium-bodied over light and full-bodied wines. At younger ages, sweeter wines are preferred. Fruit flavors and aromas as well as wood/oak and a matching mouthfeel are best appreciated among females. More males, on the other hand, prefer the aged character of wine.[23]

Of all extrinsic factors, the role of price merits being mentioned specifically. Ponder the question, "Are expensive wines extremely good or are extremely good wines expensive?" It is the gastronomy version of "what was first, the chicken or the egg?" In answering the question, several elements can be mentioned. The first is that there should always be a relation between the intrinsic and extrinsic qualities. In other words, the promise of the label, brand, qualification, or whatever needs to be confirmed by the wine in the glass. Marketing can do a lot, but it cannot make gold from iron. It can, however, give a gold finish to the iron that may fool some people at least for some time. Therefore, objective wine journalists have an important function to inform the consumers about the good wines and about the ones that just have been marketed to impress. But this is only a part of the story. Even if expensive wines are indeed very good, this still does not mean that the opposite is true (cheap wines are not good) or that such great wines are always good.

The gastronomy way of looking at the quality of wines puts them in the perspective of fitness for use: an inexpensive wine that fits the situation will give more pleasure than an expensive wine that does not fit.

You can destroy any wine—great, good, or just acceptable—with the wrong food. On the other hand, the fact that wines are inexpensive may have little to do with their intrinsic quality. They just happen to be not fashionable or to come from a small region that does not have the funds or the vision to do proper marketing. Some inexpensive wines are in fact the pearls of the wine world; they need confident and independent buyers with an open mind to find them.

SERVING

After all that has been done to make the wine, we have come to the final stage: opening and serving the bottle. Clearly, this last step is out of control of the winemaker. He may just hope that everything that he has put into the bottle is appreciated and liked. The consumer can do many things wrong. This whole book is about being aware of giving the wine the place—which includes the meal—that fits its properties. In this section, we will limit ourselves to the role of the serving glass, serving temperature, and the decanting of flavor.

The glass that is used to serve the wine may have quite some influence on the wine. To start with, glasses should be clean and free from odors. This may sound logical but, unfortunately in daily practice, it happens that the flavor of a wine is negatively influenced by the glass. Next, it is often suggested that the glass shape has a direct impact on wine aroma. Wine glasses are found in many shapes and sizes, and there are also glasses that have been designed specifically for grape varieties and wine types. Although suggestion and marketing do play a role, research shows that indeed the shape of the glass seems to influence the perception of wine and especially the perception of odors.[24]

Temperature is the next important element of serving; its influence on flavor would deserve much more attention. Temperature has an impact on mouthfeel and flavor richness. If wines are served on the cool side (at 12°C or lower) the mouthfeel will be more contracting. The flavor richness is also likely to be lower, as odorants are not as volatile. Above 20°C, freshness is lost and not the most attractive volatiles are released. Ideal serving temperatures of many wines are as follows:

- Contracting, fresh wines: 8°C–12°C
- Balanced wines, slightly coating: 12°C–14°C
- Wines with a high flavor richness and complexity: 14°C–18°C

Note that there is no distinction between red and white wines as far as the flavor theory is concerned. This also implies that, in daily practice, wines are often served at a temperature that is not perfect (to put it mildly). In general, white wines are often served too cold and red wines too warm. Furthermore, the concept of room temperature needs reconsideration. Perhaps rooms were once at 18°C, but these days the ambient temperature is more like 21°C or higher. This is too high by any standard. Restaurants are not a good storage place for wines, and the idea to put wines on a cheese trolley and pour from these bottles should also be reviewed. The use of a wine refrigerator or a dedicated storage room with climate control is preferred, if there is not a perfect cellar around.

The third element of serving has everything to do with storage capacity. If there is a wine collection, then you are able to keep the wine and drink it when it is at its peak. If not, it is likely that the wine is drunk too young, before it has reached its optimum. This can partly be solved by giving the wine the oxygen that it needs. Airing the wine by decanting it once or several times is the elegant way to help flavors develop. Traditionally, decanting was done with older wines to prevent sediments from coming into the glass. Aesthetically, this may be defendable, but from a flavor point of view it is risky. The sediment is what is used to protect the wine from aging too quickly. Decanting an old wine is therefore only advisable if the wine is still strong enough to withstand the extra oxygenation.

BEER AND WINE COMPARISON

Now that we have come to an end we may safely conclude that wine and beer, although both extremely popular in gastronomy and seemingly rather similar, are in fact very different. The differences are summarized in Table 5.4.

TABLE 5.4
Fundamental Differences between Beer and Wine

Wine	Beer
Natural juice from grapes	Water is added
Sugars directly available	Sugars indirectly available from starch after malting and mashing
Use of available or cultured yeasts	(Big) Breweries have their own specific yeast— company secret
Only natural aromas from grapes and fermentation	Freedom in adding aromas, herbs, and spices
Bitter not very prominent	Prominent vegetal bitters from hops
Variation in sweetness	Never very sweet
No CO_2 (except for sparkling wines)	Always CO_2
Strong regulation on wine types and regions	Less regulation, except for the use of basic materials
Process more artistic and closer to nature	Technical and sophisticated process, closer to industry
Focus on regions and wine types, few worldwide brands	Dominated by brands and marketing

SUMMARY

The flavor of beverages is in some ways more difficult to grasp than the flavor of foods. Ingredients are less visible, and we have always been led to believe that the labels of bottles give adequate information on flavor. They do not. The role of the brewer or winemaker is in many instances greater than the region, origin, or grape variety (for wine). Beer is different from wine—not only as a beverage but also in the philosophy of making. Beer is a more technological, industrial product with a strong dominance of commercial brands. The world of wine is more artistic, is less dominated by global brands, and is less uniform. This does not mean that the wine world is not interested in technology or that all beers are the same. In either product group, there is a lot of development. The wine world is quickly catching up in getting to know the technology behind the product, and the beer community is alive with more and more small-scale, boutique breweries that excel in individuality.

Water is an important ingredient in beverages—even in wine, although indirectly through rootstocks of vines and via grapes. When water is the direct base, it is crucial to know about and pay attention to its composition. No matter if you drink it as such or use it to make tea, coffee, or beer, the characteristics of the water determine the flavor of the brew. Rather than what was often thought in the past, a certain amount of minerality is preferred. Water that is too soft leads to low flavor intensity. Water that is too hard will have a negative influence on the brew, in most cases.

Wine is arguably the most intriguing of the beverages. This may be because man has less influence. You need top quality grapes to make very good wines. And however people can exercise their influence, we have no control over the weather. Before and after harvest there is a lot that a man can do to influence the flavor of the wine. These possibilities range from agricultural practices to pruning to selecting the harvest date to controlling the fermentation to aging and bottling the wine. The devil is in the details, they say. Well, this is certainly true in winemaking.

Finally, there is us, the consumer. We have to try to understand what has been done to create the flavor and try to describe the flavor of beverages in mouthfeel and flavor richness. We can use that information in making a choice, such as when and with what we are going to drink that beverage.

NOTES

1. This was the *Décret relatif aux poids et aux mesures* [Decree relating to the weights and measurements] of April 7, 1795.
2. www.fda.gov/NewsEvents/Testimony/ucm170932.htm; retrieved July 2012
3. Van der Aa, N. G. F. M. (2003). *Classification of mineral water types and comparison with drinking water standards*. Environmental Geology 44:554–63.
4. In this section we will limit ourselves to real tea and leave the aromatized fruit teas out of the discussion, however popular they may be.
5. See, for instance, www.lallemand.com.
6. Fleet, G. H. (2008). *Wine yeasts for the future*. Federation of European Microbiological Societies, Yeast Res 8:979–95.
7. Romano, P., et al. (2003). *Function of yeast species and strains in wine flavour.* International Journal of Food Microbiology 86:169–80.
8. Gayevskiy, V., Goddard, M. R (2012). *Geographic delineations of yeast communities and populations associated with vines and wines in New Zealand.* The ISME Journal 6:1281–90.
9. Pretorius, I. S. (2000). *Tailoring wine yeast for the new millennium: novel approaches to the ancient art of winemaking.* Yeast 16:675–729.
10. Molina, A. M., et al. (2007). *Influence of wine fermentation temperature on the synthesis of yeast-derived volatile aroma compounds.* Applied Microbiology and Biotechnology 77(3):675–87.

11. Charoenchai, Ch., et al. (1998). *Effects of temperature, pH, and sugar concentration on the growth rates and cell biomass of wine yeasts.* Am. J. Enol. Vitic 49(3):283–8.

12. Torija, M. J., et al. (2002). *Effects of fermentation temperature on the strain population of Saccharomyces cerevisiae.* International Journal of Food Microbiology 80:47–53.

13. Robinson, J. (ed) (2006). *The Oxford companion to wine.* Third Edition. London: Oxford University Press.

14. Hiromichi, K., et al. (1989). *Role of free amino acids and peptides in food taste.* In *Flavor chemistry* (pp. 158–74). Washington, DC: American Chemical Society.

15. Klosse, P. R. (2004). *The concept of flavor styles to classify flavours.* Academic Thesis, Maastricht University | Academie voor Gastronomie, Hoog Soeren, the Netherlands.

16. Moreno-Arribas, M. V., et al. (2002). *Analytical methods for the characterization of proteins and peptides in wines.* Analytica Chimica Acta 458:63–75.

17. Fleet, G. H. (2008). *Wine yeasts for the future.* Federation of European Microbiological Societies, Yeast Res 8:979–95.

18. Charpentier, C., et al. (2005). *Release of nucleotides and nucleosides during yeast autolysis: kinetics and potential impact on flavor.* J. Agric. Food Chem. 53(8):3000–7.

19. Bauer, R., Dicks, L. M. T. (2004). *Control of malolactic fermentation in wine. A review.* S. Afr. J. Enol. Vitic. 25(2):74–88.

20. Liu, S.-Q. (2002). *Malolactic fermentation in wine—beyond deacidification.* Journal of Applied Microbiology 92(4):589–601.

21. Pérez-Prieto, L. J. (2003). *Extraction and formation dynamic of oak-related volatile compounds from different volume barrels to wine and their behavior during bottle storage.* J. Agric. Food Chem. 51(18):5444–9.

22. Bisson, L. F., et al. (2002). *The present and future of the international wine industry.* Nature 418:696–9.

23. Bruwer, J., Saliba, A., Miller, B. (2011). *Consumer behaviour and sensory preference differences: implications for wine product marketing.* Journal of Consumer Marketing 28(1):5–18.

24. Hummel, T., et al. (2003). *Effects of the form of glasses on the perception of wine flavors: a study in untrained subjects.* Appetite 41(2):197–202.

CHAPTER 6

Matching Foods and Beverages: The Fundamentals

TASTE!

Combining wine and food has a sophisticated appeal; it sounds like something that worries the rich and famous or privileged people who have access to special wines and gourmet foods. However, I dare to say that all people have experienced how the flavor of one substance can be influenced by another. Moreover, most of them have done so quite often and already early in the morning. They got up, took a shower, brushed their teeth, dressed, and moved to the kitchen to grab something to eat or drink. Have you ever experienced how a slice of apple, a cup of coffee, or a glass of orange juice tastes just after you brushed your teeth? It tastes terrible, and for many people it is a reason to brush their teeth after breakfast.

At the breakfast table, new flavor pairing problems may arise, for instance, if one were to taste honey and fresh orange juice simultaneously.

If the honey is tasted first, the orange juice is likely to be perceived as being quite acidic. The other way around there is no problem. And how about chocolate paste and green tea, does that fit? And what to eat first, a slice of apple or a banana? In other words, finding good combinations of flavor is always a challenge and certainly not limited to wines or to the dinner table or to just a small group of privileged people.

It is often suggested that the mission of matching foods and wines should be to enhance each other. Or, to phrase this differently, a certain dish should make a wine taste better, or a wine should be there to boost the flavor of the food. Although this certainly happens occasionally, it goes way too far to charge wines or foods with such a responsibility. Furthermore, if this elevated principle should be true, it should not only apply to wines but to all other beverages as well. A normal, and indeed less sophisticated, starting point in finding good combinations would be that foods and beverages have a quality of their own. Chefs, brewers, winemakers, and other producers are flavor makers. They have done their work—hopefully as well as they could—and achieved a certain result. It would be rather unsatisfactory if they were dependent on an outside agent to make the flavor of their product complete. This simply is not the case.

This is similar to a good painting and a wall. The painting is independent of the wall and needs to be intrinsically interesting. Above all, the wall should be functional; the painting must fit and the wall should not fall apart or impair the painting in some way. However, on the right wall with proper lighting, the painting may shine and show its real quality. Either the food or the beverage can be like the wall. They are mutually related and should at least have a functional fit. The gastronome is content if the food and the beverage have things in common and if their identity stays intact—in other words, if the work of the flavor makers is respected. The gastronome dreads negative effects: matches that diminish the flavor of either the food or the beverage, or even of both. And indeed, we cherish the occasions where flavors are boosted. Hence, to limit the risk of deception as much as possible, it is imperative to pay attention to the factors that might impair the match. It is just as with any other sport: you prepare yourself as well as possible to have a chance to win; you do not want to lose but accept a draw.

OLD GUIDELINES

Guidelines have been suggested to help people with their wine and food choices. The traditional ones are generally known: white wines go with fish; red wines with meat; drink first white, then red; and young before old. These "rules" sound rather logical, and they are certainly easy to apply. However, have you ever really contemplated what you suggest when you say that white wine goes with fish? You are actually saying that all white wines are the same and, furthermore, that all fish are the same no matter how they are prepared or whatever is served with them. Clearly, this is nonsense. White wines can be very different, and this has always been the case. In addition, a salmon is not a tuna, and a poached cut of salmon with vinaigrette has never tasted the same as a fried piece with tomato ketchup. We could elaborate on the value of the other guidelines but would quickly come to similar conclusions. Indeed, there are many white wines that go with meat—situations where it is better to serve red wines before whites and to have the older bottle before a younger one.

The problem with the traditional guidelines is that they are based on elements that do not say much about flavor, such as color, age, or origin. One of the first books that challenged traditional wine and food recommendations was *Red Wine with Fish. The New Art of Matching Wine with Food* by Rosengarten and Wesson.[1] In a wonderful, ironic way, they state: "To become an expert on matching wine and food, there are only three things you need to know: 1. The taste of every wine in the world, from every vintage; 2. the taste of every food in the world, from all producers; 3. how every wine and every food will taste together in every circumstance."

They conclude: "There are no experts on matching wine and food." Although I cannot fully agree with this conclusion, the irony is clear. Indeed, we do need to taste the food and the wine or whatever beverage. In finding good combinations, we need to focus on the intrinsic factors. Without knowing more about those, we can only go by the extrinsic clues, such as the label of a bottle or the description of a dish. However, looks can be deceiving. There are so many variables that general claims based on the extrinsic components are almost impossible. In real life, the influence of the chef, the brewer, or the winemaker is too great. There is only one solution: taste!

In this chapter, we will introduce new and useful guidelines that help in finding good matches of foods and beverages based on the intrinsic components. After the introduction of the new guidelines, we will elaborate on matters that influence finding good combinations; after that, some specific issues are discussed, such as combinations that involve cheese, beer, and tea. Clearly, these guidelines are based on observations in practice.

NEW GUIDELINES

It is futile to even try to taste all the foods and beverages of the world, and it is certainly not an efficient way to become an expert in the field. Furthermore, foods and beverages are fundamentally different, which makes it impossible to formulate guidelines on the product level. Successful recommendations need a common ground. Flavor is what foods and beverages have in common. Therefore, the descriptors of flavor, the universal flavor factors, are now used to solve the problem.

In order to find good matches, we need to have a notion of the flavor profile. Foods and beverages do not just accidentally have a certain mouthfeel and flavor richness. Their flavor profile is the result of basic qualities and of what has been done and/or added. What will the mouthfeel be: neutral, dry, contracting, or coating? And what about the flavor richness? What do we expect about the flavor intensity and the flavor type; will the fresh or the ripe flavor tones predominate? To be able to assess all that, it helps to know the origin of the food or the beverage and how it has been prepared. Understanding the elements that contribute to flavor greatly helps. The previous chapters have shown that. This means that the true gastronome must be an expert in his field. The more he knows about what constitutes flavor, the better he is able to assess it and to find good combinations of flavors.

Modern guidelines on combining foods and beverages follow quite log-ically from their flavor profiles. Overall, successful combinations can be expected when the profiles of food and beverage match or at least bear a strong resemblance in the most important elements. They are formulated as follows:

- Mouthfeel needs to match: that is, contracting fits contracting, coating fits coating
- Flavor intensity needs to match
- Flavor style needs to match: fresh fits fresh, ripe fits ripe

As far as the order is concerned, the following may be added:

- Contracting before coating
- Fresh before ripe
- Low intensity before high intensity

This is the essence of matching foods and beverages. The basic premise is logical and easy to understand. Application will take some practice. Fortunately, eating and drinking are daily activities and, therefore, there are plenty of opportunities to test the sugges-tions. Experience shows that it is easier than you first may think. Learning by doing is not too bad in this respect and, as always, practice makes perfect.

THE DOMINANT FACTOR

The universal flavor factors represent the side of flavor that can be assessed rather objectively, particularly when more people are involved. Guidelines for flavor matches have also now been formu-lated on the same basis. The logical consequence must be that com-bining foods and beverages has a degree of objectivity. After 20 years of experience, I dare to say that this is indeed true. Chances are that combinations of foods and beverages are preferred more often when the flavor profiles match. Or, a certain wine that matches the food will be liked more—and by more people—than a wine that does not match. Please note that this is not a guarantee. And it cannot be. After all, the final judgment on preference is personal. Nevertheless, raising the odds of liking is already very valuable, especially in commercial situations.

The rest of this chapter is filled with observations and reflections that should help one to find interesting matches of foods and beverages. However complicated it sometimes may seem there is always the basic premise to fall back upon: identify the dominant factor. This is the key factor that primarily characterizes the flavor. The food and beverage should at least be on par on this point. Imagine a green salad with some tomatoes, capers, cucumber, goat cheese, and pine nuts. Each of these ingredients has a specific character. Nevertheless, the dominant factors of this salad are the fresh tones and the contracting mouthfeel. Therefore, first and foremost, the appropriate wine needs to have fresh tones and to be on the contracting side.

If, in another case, the flavor profile is dominated by a high flavor intensity, then that is the factor to watch. If the beverage does not equal this intensity, it has no chance to show its flavors because it will be dominated by the dish. The same applies the other way around. In gastronomic reality, there will practically always be dominant factors. Other factors might be present, but not on center stage. They come into play after the dominant ones are dealt with.

Now, everybody knows that only the tip of an iceberg is visible above sea level. The rest is invisible but undeniably there. Something similar applies to gastronomy. There is more to it than meets the eye. Therefore, let us see what is below the surface and elaborate on the issues that have an impact on combining foods and beverages. In our discussion, we will distinguish between influences that have a gustatory origin and influences that involve mouthfeel. The distinction between these two effects is somewhat blurred. To illustrate, acidity and salty are gustatory in origin but have an impact on mouthfeel—just as sweet or umami, only differently.

GUSTATORY EFFECTS IN MATCHING FOODS AND BEVERAGES

THE RICE TEST

Sweet, salty, sour, bitter, and umami are the established basic tastes for now. Most of them can easily be added to a dish, even at the last moment. There is no doubt that each of these elements has a major impact on the flavor of our foods and consequently on their combination with a beverage. This can be experienced with a rather simple, yet elucidating, test: the rice test. Boil simple long grain rice with a neutral character. Let it cool down, divide it into four portions, and then add white wine vinegar, salt, sugar, and a mix of Tabasco and chili peppers, respectively. Bitter is not a part of this lineup because of practical reasons—which bitter to use and how to add bitterness to the rice in an easy way.

Also select four distinctly different wines: two whites and two reds. For the whites, select one that is fresh and contracting (e.g., a classic Sauvignon blanc) and one that is full-bodied and on the coating side (e.g., a popular wooded Chardonnay from a warm climate). For the reds, one preferably should be young and tannic (e.g., a classic Bordeaux), and the other should have no tannins but rather be light and fruity (e.g., a light Burgundy or Beaujolais). Alternatively, a mild white one with some residual sugar is also interesting to use—for example, a German Riesling *trocken* (if not too acidic) or *halbtrocken* (if not too sweet) or Vouvray Tendre.

Start by first tasting the acidic rice with each of the four wines, then the salty rice, after that the sweet rice, and finish by having the spicy rice with the wines. Focus on the effect that the basic flavors have on wine, not on the way the wine interferes with the rice. It is quite an effective way to experience the influence of the flavors and to understand their impact. The following is what normally can be expected.

ACIDITY

In general: Acidity does not create many problems with wines because wines have acidity as well.

White wines: Overall white wines react better than the red ones. Wines with some sweetness will taste sweeter. Wines with a high acidity may react negatively as acidity adds up. It can become too much. The full-bodied types often react in a positive way; their balance seems to get better.

Red wines: The fruity types will appear to be more full-bodied; their fruitiness is enhanced. The tannic types tend to taste harder.

SALTY

In general: Salt is a flavor enhancer and can cause problems. This is largely dependent on the quantity used.

White wines: White wines are mostly not impaired. The full-bodied types may gain some fattiness. The biggest problem is to be expected with wines that have a lot of alcohol: they may appear even more alcoholic and lose their fruitiness.

Red wines: The lighter and fruity types may lose some of their character; the tannic types tend to get harder.

SWEET

In general: The deciding factor is if the wine has some degree of sweetness. Therefore, it is useful to know the grams of residual sugar of wines. Sweetness in food has a tendency to accentuate acidity.

White wines: If the wine has some sweetness, the combination will be rather positive if the sweetness of the dish matches that of the wine. In the best cases the wine is just a touch sweeter. If dishes are sweeter than the wine, the wine seems tart.

Red wines: Red wines are mostly dry and, therefore, most of them have a problem with the use of sugar in dishes.

Spicy

In general: What we call spicy is actually an irritation of the endings of the trigeminal nerve. In other words, spicy is not a gustatory flavor; it is a matter of mouthfeel. The effect of spicy is comparable to a little wound on your hand that comes in contact with acidity; it hurts even more.

White wines: The best matches are with those wines having some sweetness. Wines that are high in acidity are to be avoided. Milder types that are low in acidity and higher in alcohol are a better choice. They will generally lose a bit of their nuances as papillae are numbed to some degree.

Red wines: Tannic red wines, especially the Bordeaux types, may react surprisingly well with spicy flavors: a possible explanation is that tannins have less impact on saliva, giving room for other elements of the wine to appear. The fruity red wines react like the white ones: they seem to lose nuances.

Conclusion

Some types of wine are more vulnerable than others. The classic dry white wines especially require particular care; they are easily out of balance. The tannic reds also call for attention but may surprise. The fruity style reds are easier to use, but the level of acidity needs to be watched. The type that is the least specific and rather adaptive to gustative flavors is the full-bodied white. These whites generally perform well, which explains their commercial success from a gastronomic point of view. See Table 6.1.

Acid Observations

With regard to gustatory flavors, acidity is not much of a problem in combination with wines and other beverages. The logical explanation is that all wines have a degree of acidity. Therefore, as far as wine is concerned, acidity in the food is not a substance that cannot be handled. Nevertheless, there are some aspects of acidity to observe when matching foods and beverages.

TABLE 6.1
Reactions of Basic Flavors on Wine Types

	Classic Dry White (contracting, fresh flavor tones)	Full-Bodied Whites (slightly coating, ripe flavor tones)	Light and Fruity Red (contracting, fresh red fruit flavors)	Classic Tannic Red (contracting, dry, ripe flavor tones)
Sour	Freshness and acidity enhanced	Balance is enhanced	Fruitiness is enhanced	Harder, but balance stays
Salty	Loses some fruitiness	Fattiness is enhanced	Some fruitiness may get lost	Harder, but balance stays
Sweet	Acidity is enhanced, tartness	Acidity is enhanced	Acidity is enhanced	Harder, balance is lost
Spicy	Freshness and acidity enhanced	Nuances are suppressed	Nuances are suppressed	Softer, fruitiness appears

These are general conclusions. Individual wines may react differently and surprise you. That is a part of the gastronomic fun.

First, it is important to know that acids add up. Yes, it can be handled, but adding extra sourness to something that is already acidic can become too much.

Second, wines can have a variety of acids. Most common in wines are malic, lactic, and tartaric acid. Malic acid is the strongest of the three. Malic comes from the Latin word *malum*, meaning apple. The flavor of malic acid resembles that of green apples. Some grape varieties such as the Carignan, Barbera, and Sylvaner are rich in malic acid, and many cool climate grape varieties are also known to have significant amounts of malic acid. During ripening, this acid is metabolized to some degree, especially in warmer climates. This implies that malic acid can be expected to be more obvious in grapes of cooler climates and years. The winemaker has a tool to reduce the amount of malic acid: malolactic fermentation, often abbreviated as MLF. In this process, malic acid is converted into lactic acid, which is softer than malic acid. Consequently, a wine with little or no malic acid will taste less sour than a wine with a higher content. Most of the red wines undergo malolactic fermentation. Wines of grape varieties such as Riesling, Chenin blanc, and Sauvignon often contain malic acid because they are likely to develop a specific buttery flavor (diacetyl) that does not fit the primary flavors that characterize these wines. On the other hand, this buttery flavor is preferred in Chardonnay wines. In conclusion, the combination of dishes with high acidity and white wines from cooler regions needs to be observed closely. It can be a perfect match, but if the acidity levels get too high, the combination turns bad.

Third, we need to consider this from the other side as well: the dish. The chef is free to use whatever he wants, and many other types of acidity are at his disposal. Not all of those are friends of wine. Citric acid, found in many fruits, is one of them. Therefore, the use of fresh fruits in dishes can be quite a problem for many wines, especially the ones that are not highly acidic. They will lose their specific character and become bland. Interestingly, there are quite a few classical dishes where fruits are used. In most of these cases, the fruit is used in the preparation and loses much of its acidity. Alternatively, canned fruits may be used because they lose their acidity in the process of sterilization. Fruit character and acidic profile changes when heated; it is the raw slice of lemon, orange, or kiwi on foods that is to be feared.

To finish our observations, it is good to note that acidity is an interesting element in dishes. It may not be the most popular, and its importance appears to be somewhat overlooked these days. In modern convenience foods, sweetness is predominant. However, in the process of ameliorating combinations with beverages, adding a touch of acidity to dishes can do wonders. If there is a sauce, adding just a bit of the wine that is served with it will improve the match.

SWEET CONNECTIONS

Sweet and sour are often considered each other's opposites, but in fact they are not; they are even quite different. Nevertheless, sugars have the capacity to mask acidity, and many chefs add sugar to their dishes for that reason. A familiar example is vinaigrette. By now, many people seem to believe that sugar water or honey is even a standard part of the recipe. Traditionally, they certainly are not: vinaigrette is a mixture of vinegar and oil (with some mustard and seasoning). The balance is found in the right mix of the two and is dependent on the sharpness of the acidity and the viscosity of the oil. If the vinaigrette were too acidic, the normal reaction would be to add more oil. Adding sugar can do the trick as well, but that is another type of solution. The sugar does indeed make the vinaigrette taste softer, but the effect on wine will be markedly different. This is important to know. The rice test shows that there is a significant chance that the use of sugar in dishes has a negative effect on wines, especially on the ones that have no residual sugar at all. This can even be the case when the sugar is hardly perceived.

What can be said about dishes applies to wines as well: beverages can be perceived to be dry even though they have a certain level of sweetness. In Germany, a wine labeled as "trocken" may contain up to 9 grams of sugar/liter. Champagne that is called "brut" may even contain up to 12 grams of sugar/liter. In both cases, the wines will definitely be perceived as dry, despite the considerable amount of sweetness. Many modern wines have some residual sugar that also goes unnoticed. In all of these situations, these wines will handle the use of sugar in a dish more easily; they have some kind of a buffer to absorb the sweetness.

In the earlier example, sugar is added in some way or other. Some vegetables have a sweetness of their own, and this can be enhanced or developed in a cooking process. Think of roasting or grilling bell peppers, mainly the red ones. When the skin is blackened, it can easily be removed; the flesh will taste rather sweet. Onions can be stewed with butter and develop their natural sweetness, just as can beetroot, to mention a few examples. Interestingly, this sweetness does not interfere with wines as the added sugars do. On the contrary, beetroot and many red wines are natural allies.

Up to now, we have discussed dry wines or wines that seem dry but that have a certain degree of sweetness. How about the wines that really taste sweet? The general rule is that these are best with sweet dishes. The thing to watch is the level of sweetness: preferably the amount of sugar in the beverage is just a bit higher than the dish. If the food is sweeter than the wine, the wine will be perceived as more acidic; if the wine is too sweet, it will dominate the dish. There are exceptions to this rule. Sweet wines go well with spicy foods and with foods that are extremely coating and rich in flavor, such as a pâté de foie gras and some cheeses. The level of sweetness needs to be in line with the flavor richness (e.g., with spicy foods "medium dry" will be a good choice, while Sauternes is known to go well with goose liver or Roquefort).

FLAVOR STYLES AND DESCRIPTIONS

Sweet and acidic are frequently related to certain flavor styles. Sweet is often associated with ripe flavor tones, and sour and fresh are also strongly associated, however not always. Some practical examples illustrate this. A grape variety such as the Muscat has predominantly fresh flavor tones even if a fortified sweet wine such as a Muscat de Beaumes de Venise or a Muscat from Samos is made from it. Consequently, these wines are best with dishes with comparable, fresh aromas. Desserts with fresh fruits (apple, pineapple, citrus, etc.) work better than desserts with ripe flavor tones such as chocolate or nuts. In a similar way, the specific aromas of the Gewürztraminer (spices, flowery, lychee) also fit similar characteristic aromas in food. These wines are an interesting choice with spicy food, and not only because they are often on the sweet side. Coriander leaves (cilantro) match especially well with Gewürztraminer.

There is also another side to contemplate. Descriptions of wines can be very poetic, mentioning all kinds of elements, fruits, and specific flowers. Read these actual descriptions and imagine dishes that would match—before you continue reading.

> A nose which opens out within an hour or two to reveal
> the aromas of chalky blackberries, beautifully vibrant
> and pure, with a fresh and slightly leafy note. A lovely,
> cool and fresh style on the palate, sappy and with

well-defined juicy fruit, full of cherry and blackberry. This wine has a fine texture, savory and dry. Through the middle it has a fine, firmly composed structure, which spirals down into a pure drop of fruit at the very end. In the finish, a little flourish of tannins gives some grip. A really impressive, finely defined wine.[2]

On the nose an elegant layer of creamed fruit here, reserved rather than exuberant. Slightly smoky, but very well defined and composed. The palate has a fine layer of creamed fruit which slowly unfolds on the palate, and it covers the alcohol very well, as well as the building tannin. Beautifully sweet, rich, expansive but also well contained within the frame of acidity. This is an exciting and very vibrant wine. Very stylish, composed and elegant, with a sleek and seamless composition, yet with great depth.[3]

Now, what are we going to have with these wines? What color would they have in the first place? The first description points in the direction of red or rosé by mentioning cherry and blackberry. In the second, there is not much to go on; the only word that suggests at least something about color is tannin. It could indicate that it is a red wine, but white wines can be tannic as well, especially if they have been matured in wood. However, there are no other references that indicate the use of wood. Furthermore, words such as *creamed*, *sweet*, *rich*, and even *acidity* are easily associated with white wines. The use of the word *vibrant* supports this suggestion even more. Could it be Sauternes or any other rich and sweet wine that is aged in a barrel?

Now about the ideal foods: the first wine is likely to be young, fruity, and juicy; red and black fruits are abundant. Would one idea be to go get a basket of these fruits? The words *dry* and *tannins* are worrying; sweetness in the dish can be problem. Are we having a steak with cherries on top and a blackberry sauce? The second one gives less reason to worry. The description suggests that it is a perfect wine with an apple pie with a vanilla sauce, does it not?

As long as we only have the words to go by, these proposals would certainly be taken seriously. As soon as we know the wines, this is likely to change. In fact, the first description is of a 2007 Brouilly, tasted in 2010, which would normally be considered rather old for a Beaujolais wine. It would be astounding if it were as young and lively as is suggested. Any dish in which red fruits are used is very likely to kill the wine. The second description is of one of the most prestigious and expensive red wines of the world, 2009 Lafite Rothschild (Pauillac, Bordeaux), which is surely about the last wine to have with an apple pie.

Therefore, do not be fooled by descriptions. Fruit flavors that are identified in wines are just verbal impressions of an aroma; using the actual full and complete fruits or flowers in a dish is asking for disappointments. The same must be said from words such as *sweet*. The Lafite may be perceived as full-bodied, concentrated, and rich, and yet even a small amount of sugar in a dish would be perfectly capable of degrading the majestic Lafite. It has been said before: tasting is imperative.[4]

MOUTHFEEL EFFECTS IN MATCHING FOODS AND BEVERAGES

TYPES OF COATING

When the flavor factors were introduced, four kinds of mouthfeel were distinguished: neutral, dry, contracting, and coating. Distinguishing differences in mouthfeel is important with regard to combination with beverages. It all starts with the earlier mentioned concepts of *hydrophobic* and *hydrophilic*, or rather the impossibility of fat and water mixing. Obviously, all beverages mainly consist of water. By nature, they have problems mingling with fat. This principally means that problems are to be expected with fatty sauces or complete dishes. The amount of fat is important. Smaller amounts are easily absorbed and not very problematic. The richer in fat the sauce or the dish becomes, the more coating the wine needs to be.

Occasionally you hear somebody suggest the contrary: fatty foods need acidic wines to break down the fats. Although this is technically true, there is no reason whatsoever to let that process start in the mouth. Our digestive system, and particularly the acid of our stomach, is much better equipped to do this. As mentioned before, a coating dish will enhance acidity in a beverage. The larger the gap in mouthfeel between beverage and food, the less pleasant the combination will be.

If indeed a wine is too contracting for a certain dish, it is important to note that adding fresh aromas is not the solution to the problem. Therefore, it does not help to add, for example, tarragon, dill, or parsley to a sauce that is too coating for the wine. Freshness is a part of flavor richness and cannot solve a mouthfeel problem.

Sauces are a major (and simple) contributor to mouthfeel in dishes. Sauces vary from contracting to coating. In the contracting sauces, acidity is mostly dominant. In coating, fat, proteins, and carbohydrates can be the source. Of these, sauces that are rich in fats are likely to give the most problems in combination with wines. Sauces enriched with butter or oil are examples (e.g., the so-called *beurre blanc*). Other coating sauces are less troublesome. The sauces that are the result of a careful reduction, such as a fond or demi-glace, get their jelly-like texture from the decomposition of proteins. Nor will sauces that are thickened by starch or another carbohydrate cause many problems with wines. The explanation is that the last two families of sauces are hydrophilic; their composition allows beverages to mix more easily than those sauces that are coating because of their fat content.

FIZZY SECRETS

Carbon dioxide, CO_2, is a special class of mouthfeel on its own. Evidently, the gas gives sodas, beer, and sparkling wines its specific character. Carbonation strongly influences mouthfeel and belongs to the contracting influences. Nevertheless, research has uncovered that its flavor is not purely trigeminal or somatosensory; the gustatory system is also involved in the registration of CO_2.[5]

Although sparkling beverages are clearly contracting, and therefore at their best with contracting dishes, they also have a special quality. They can be combined with some coating foods that are otherwise hard to match. Soft boiled, or poached, running eggs are among the examples. Most "normal" wines are in deep trouble, but the fizz spikes right through the coating layer and is not hurt by doing so. Beer and sparkling wine act similarly in this respect. Besides these coating proteins, carbon dioxide fits quite well with the coating starches as well. Try having a sparkling wine with a risotto, for instance. It is coating; nevertheless, it is likely that you will not be disappointed. Likewise, effervescent beverages can be interesting to have with cheeses (as will be discussed in the section "Say Cheese" in this chapter).

In the world of sparkling wines, it is useful to know how the wine is made or, more specifically, how long the second fermentation in the bottle has taken—the longer, the better. The CO_2 is developed and captured in the bottle during the process. After the fermentation, the now inactive yeasts (the lees) have another function that is less evident. It is called autolysis and refers to an enzymatic process that frees glutamate; umami is formed. The result is that flavor richness is enhanced. This characterizes the highest category of champagne, the *millesimés* and *cuvée prestige*. Their maturation on the lees often takes more than five years. The less expensive sparkling wines get their carbonation in a tank or the CO_2 is added. Evidently, there will be no effect on flavor richness in those cases.

THE DRY RUN

Dryness is yet another class of mouthfeel and different from acidity. Acidity is more a gustatory flavor, whereas astringency is more in the trigeminal world. It is caused either by absorption of saliva or by a change in the composition of saliva. Starches in the form of bread, toast, crackers, biscuits, rice, and so on are good at absorbing saliva. The logical and indisputable consequence is that the mouth gets dry. The use of drying accompaniments is functional. Coating influences can be compensated by dryness. Think of how toast or biscuits reduce the perception of fat in the case of smoked salmon, pâté, or fatty cheese. Likewise, the perception of sweetness is reduced if marmalade is spread on toast. Rice has a positive impact on the impact of spiciness (as noted in the section "The Rice Test" in this chapter).

Mouthfeel can also run dry due to a change in the composition of saliva. This is likely to be the result of the consumption of tannins.

Although the composition and flow rate of saliva varies among individuals, it always contains many proteins that are rich in proline. Together with some other substances, this gives saliva its viscosity. Tannins interact with these proteins; the result is that saliva loses its lubrication or at least some of it. Tannins have a drying effect.[6] Astringency is the proper word to use in this respect; the dryness is what distinguishes it from bitter. In other words, like acidity, bitter is more a gustatory flavor; the trigeminal nerve is also involved in the perception of astringency.

Astringency is what characterizes the flavor of young red wines because of their tannins. From a wider perspective, dark chocolate, green tea, and coffee are other examples of astringency. The higher the cocoa content of a chocolate bar, the greener the tea, or the stronger the coffee, the dryer it gets. This distinction between bitterness and astringency helps us to understand the match of tannic beverages and foods. Tannic wines match well with dry or drying foods. This stands to reason. If saliva is absorbed, there is less for tannins to impact. Furthermore, astringent drinks are much less bothered by coating mouthfeel, independent of its origin. Fat and sugar can compensate for dryness in some way.[7] This may sound contradictory. The general suggestion is that contracting beverages and coating foods do not mix well. However, the different nature of astringency and of acidity causes them not to react in the same way. Astringency is not bothered too much by a coating mouthfeel from fat or sugar. Acidity is likely to be enhanced by these substances.

Hence, astringency is a gastronomic problem that can be solved. In daily life, this is reflected by offering pieces of bread in a tasting of tannic red wines. It is likely that the fruit side of the wines is enhanced by taking away their harsh tannic side. It also explains why these wines go well with roasted meat: the crust absorbs the tannins, at least partly. And if you want, it opens new doors: how about matching dark chocolate and young red wines? It works very well, particularly in the case of "new world reds" with around 14% alcohol and/or some residual sugar to match the sugar of the chocolate. On the other hand, not much pleasure is to be expected when the tannins can freely do their destructive work on proteins, as is the case with fish or meat that is not "protected" by a crust.

SPICY PROBLEMS

The first thing to say about spiciness has everything to do with adaptation. During a meal, this generally takes 15 minutes. If spiciness is a part of the daily menu, we grow accustomed to it, which explains why some individuals and even certain cultures can handle higher intensities than people who are less accustomed to gingerol, piperine, and capsaicin—substances that are the working agents of spiciness. They irritate the nerve endings of the trigeminal nervous system, which gives a tingling, hot feeling. It burns, so we say. In this respect, these compounds are the opposite of menthol, which is said to have a cooling effect.

There are many varieties of peppers. They differ in origin and especially in the amount of capsaicin that they contain. This is measured on the Scoville scale. There is a difference between chili peppers and the general pepper that most people know and use in the kitchen. Chili peppers, or chilis, are the fruits of the plant family that is appropriately called *Capsicum*. The capsaicin is mainly present in the seeds and to a lesser extent in the flesh. A well-known member of the family is the bell pepper or paprika, which has been bred to become milder and has lost practically all of its capsaicin. The spiciness of other members of the family varies from soft and subtle such as *piment d'Espelette* to extremely powerful such as the jalapeño pepper. All varieties have specific aroma compounds, especially if they are dried or smoked. The ground black or white pepper that we generally use is obtained from the fruit of the *Piperaceae*. Depending on its preparation, peppercorns are also available in green and red. Again, each type has a specific character, even within the colors. This is mentioned to illustrate that it is hard to say general things about matching spicy foods with wines. Too many variables have an impact.

However, it helps to know about some of the properties of capsaicin—besides its burn. How to extinguish the fire? Capsaicin is soluble in fatty and sweet substances and also in alcohol. This explains why red and white wines with an alcohol content of more than 13% and sweeter wines can cope with spicy foods. A part of the capsaicin is absorbed and swallowed. Technically, milk would also do the job, but that can hardly be considered a gastronomic solution. Capsaicin is not soluble in water, and acidity and carbon dioxide (CO_2) increase the burning sensation. This rules out many beers and sparkling wines as antidotes. Only those with some sweetness or those that are higher in alcohol could be considered; not much can be expected from the acidic wines that are lower in alcohol. Grape varieties such as Riesling, Sauvignon, Chenin blanc, Cabernet Sauvignon, or Pinot noir may not be the best choices in their dry versions. On the other hand, milder grape varieties such as Semillon, Viognier, Grenache blanc, Marsanne, Vermentino, Grenache, Carmenère, Merlot, and Tempranillo can do surprisingly well.

There is not much known about the specific relation between astringency and capsaicin. In tastings, tannic red wines can perform surprisingly well, possibly in combination with an alcohol content that works positively. One piece of advice in conclusion: watch the temperature. If acidity and CO_2 are in play, a temperature below 12°C will increase the burn—around 16°C would be better. This is also the best temperature for the wines that are supposed to work well.

Say Cheese

In the world of wine and food, much attention is paid to cheese. Some years ago, the general belief was that mainly red wines and particularly Port were logical and attractive matches with cheese. The origin of this assumption is found in a French gastronomic custom. Brillat-Savarin formulated it as: "a dinner which ends without cheese is like a beautiful woman with only one eye." Generally, a red wine is drunk with the main course. Hence, either the same wine is drunk with the cheese, or a higher level wine is selected to surpass the wine that was used. Evidently, Port easily does the job. The English have a tradition of ending the meal with Port as well, which is often served with their Stilton.

Interestingly, if we consider the wines preferred with cheeses in the region where the cheese originated, we get quite a different picture. White wines such as Sauvignon de Touraine or Vouvray are the preferred wine in the center of France, where many goat cheeses are made. In Alsace, Gewurztraminer is the preferred wine to go with Munster.

In the regions where bleu cheeses are made, sweet white wines such as Sauternes are the favorite. In fact, if only the cheese is considered, it is striking to see how often white wines are suggested as going well with a particular cheese.

At large, this may lead to the suggestion to have the local wine with the cheese. Although this may sound like logical and safe advice, it does not hold. The origin of Brie is close to Paris, Camembert is made in Normandy, and Gouda in Holland: three famous cheeses without a "home wine." Likewise, it is doubtful whether you will find the best matches with Parmigiano, Manchego, Emmental, Cantal, and Cheddar in their own regions.

Indeed, white wines are generally easier to match with cheese than reds. However, not all whites and with not with all cheeses. How could that be the case in the first place? Just as we cannot speak of "a wine," it is pointless to consider cheese as an entity. There are so many different kinds. The process all starts with the milk that is used—cow, goat, sheep, buffalo—they can all be used to make cheese. Moreover, even within the same origin, there are profound differences in the milk depending on the time of the year and the characteristics of the grass and of the herbs. Then consider the textural differences. Cheeses range from soft to hard, from dry to creamy. And then there are all kinds of influences on flavor: young and aged, soft and pungent, and the most important influence of all: the kind of rind and the use of blue, white, or red molds.

We pretend that universal factors are leading us in finding good matches. The same theory applies, although some cheeses can be rather choosy when it comes to beverages; combinations can be really bad. The first thing to do is determine the flavor profile of the cheese. Clearly, the soft cheeses such as Brie and Camembert are more on the coating side—the higher the fat content, the more coating. The hard cheeses are more contracting; the two important variables are salt and umami. Salt is more contracting, while umami has a coating component. Think of how a good Parmigiano, a hard cheese rich in umami, melts in your mouth and coats it. Cheeses from goat's milk are the most acidic of all. The dry, chalky, goat cheeses are particularly contracting. Flavor richness of cheeses is enhanced by aging. This is partly explained by the loss of moisture, which concentrates the flavors. That is not all; in the process of maturing, many other things happen as well. Enzymes, and in many cases molds, transform proteins and form new flavors. With the exception of specific types that are not meant for aging, young cheeses are almost by definition less rich than the fully matured ones. Such biochemical processes work particularly well with cheeses that are made from raw (nonpasteurized) milk. Champions in flavor richness are the bleu cheeses, especially Roquefort.

By and large, the same principles apply in finding good matches between cheeses and beverages. The best matches between cheeses and beverages are found if their mouthfeel and flavor richness correspond. Therefore, the combination of a Sauvignon and a dry goat cheese is not so surprising after all: both are contracting. Furthermore, white wines do indeed have the advantage compared to reds. The reds can be pleasant to drink with dry and hard cheeses, but coating reds are hard to find; whereas there are quite a few creamy cheeses, which severely impairs the use of red wines with cheese. It also explains the success of Port or sweet wines in general: their mouthfeel is coating and their flavor richness is high. This is the general perspective.

Besides still wines, other wines and beverages need to be discussed as well. The first to mention are the sparkling ones. Earlier in this chapter, in the section "Fizzy Secrets," it was mentioned that carbonation works well to spike through creamy and rich textures, which means that all kinds of bubbles are to be considered with cheeses. Besides sparkling wines, think of beer, cider, and poiré. Some even say that beer always defeats wine in combination with cheese. I doubt if this is true; certainly beer is not a safe choice. Within beers, there are many differences, and not all beers work well (see the section "Beer and Food" in this chapter). Sparkling drinks from fruits other than grapes, such as apples (cider) and pears (poiré), are more general and safer choices, especially if they have a certain sweetness. At dinner, choosing these alternatives can add extra interest because they are much lower in alcohol than many wines. Table 6.2 summarizes the types of cheese and gives some general suggestions.

TABLE 6.2
Cheeses and Beverages

Type	Characteristics	Examples	Suggestions
Soft fresh cheese	Coating, not very rich in flavor	Fromage blanc, cream cheese, mascarpone, mozzarella	Easy-drinking white or red (Beaujolais)
Goat cheese	Contracting if texture is dry, otherwise balanced. Flavor richness average	Valençay, Ste Maure, Crottin de Chavignol	Contracting dry whites (Sauvignon, Chenin blanc), white beer
Hard cheese	On the contracting side when older. More coating if younger. Flavor richness above average	Gouda, Cantal, Emmental, Conté, Cheddar, Manchego	Wooded Chardonnay or softer reds (Grenache, Merlot, Zinfandel), matured vintage Port
Very hard cheese	On the contracting side. Flavor richness high (umami)	Over-aged Gouda, Parmigiano Gran Padano	Contracting reds (Cabernet Sauvignon, Syrah, Sangiovese, Tempranillo) or young vintage Port

continued

TABLE 6.2 (Continued)
Cheeses and Beverages

Type	Characteristics	Examples	Suggestions
White rind cheese	Coating, flavor richness at least average, depending on age	Camembert, Brie, Brillat Savarin	Chardonnay, Viognier, Grenache blanc, beer
Red mold cheese	Coating, flavor richness at least average, depending on age	Munster, Reblochon, Epoisses	Rich aromatic white such as Gewurztraminer or cream sherry
Blue veined cheese	Extremely rich in flavor, contracting because of saltiness, as well as coating	Roquefort, Gorgonzola, Cabrales, Stilton	Botrytis wines, Port

GASTRONOMIC REFLECTIONS

This section is devoted to some extra thoughts on combining food and beverages. Specific issues, popular suggestions, and beliefs are being assessed—such as why it is sensible to look for harmony, why a pinch of salt can be risky, why traditional, regional choices cannot be trusted, why marriages are not made in heaven, and so forth.

Harmony and Contrast

First in line for discussion is the suggested quest for harmony. Is this really what we should strive for in gastronomy? Can contrast also lead to something good? The subject is worth elaborating.

Let us start with what harmony is. What makes flavors or flavor combinations harmonious? The essence of harmony is that parts fit well together. The antonym of harmony is chaos. This makes it easier to know what harmony is. Harmony has been studied for centuries; other sensorial worlds, such as *sound* and *color*, have come quite far in understanding what needs to be done to make elements fit well together. The same needs to be done in gastronomy. For color, great scholars such as Aristotle, Goethe, Newton, and Schopenhauer did the groundwork.[8,9] In 1722, almost a hundred years before Goethe, Rameau did that for sound. He published an innovative book on harmony in music and was the first to incorporate mathematics and philosophy in his study.[10] Color and music theorists have always sought logical structures because harmonious colors and sounds are pleasing to the eye and to the ear.

It is logical to like harmony, and this can be explained by the way our brain functions. Our brain constantly receives an enormous amount of sensory information to process. Therefore, it is set up to discard what it cannot quickly understand or organize. Coherent signals seem to stimulate the brain. Consequently, a prerequisite for liking is that the brain understands the signals. From a neurological point of view, this happens when a certain part of our brain, the orbitofrontal cortex, is stimulated in such a way that dopamine is released. This leads to a certain degree of arousal and desire.[11] This will be further addressed in Chapter 7.

In the more advanced sensory fields, vision and sound, the term *fluency* has been introduced to assess why our brain likes things. If fluency is high, signals can be quickly processed by the brain. It helps when the brain recognizes a certain structure and is able to predict what comes next; no questions are raised. It has been shown that this is an essential part of liking something. At the same time, this also gives us a better understanding of where things can go wrong. If we know exactly what is coming next, we do not get excited. Ultimately, perfect harmony will be boring because it is predictable. Therefore, elements of surprise are needed—ingredients that make things interesting. In music, a contrast is used do this. The contrast should improve the harmony, which puts the use of a contrast in its right perspective; it has to be functional. Near perfection is more interesting than sheer perfection. If something perfectly fits, you have lost any ground for discussion. Indeed, wine and food matches can be too perfect; there needs to be room for discussion.

The gastronomic road to success is searching for harmony that excites. Mouthfeel and flavor richness are the signposts that help in finding that goal. Now, suppose that a dish is extremely rich and coating. The gastronomic guidelines suggest choosing a beverage with a similar profile. The risk is that such a combination will become too heavy, even unpleasant. This is the situation where a contracting, acidic beverage can serve a purpose by restoring the balance in some way. We need to point out that this example does not work the other way around. It is not a good idea to serve a very coating wine to counterbalance an overly contracting dish. Chances are that the wine will lose much of its richness and character.

The question may be raised as to whether a beverage is meant to be used for this purpose. It becomes like an ingredient and even an essential one, for that matter. For a prepared dish, the chef can (and should!) solve this problem more easily and see to it that his dish is not too heavy to start with. When dealing with a finished product, it works to have a beverage as a contrast. An example of this is seen in the world of cheese. With the extreme ones, beer, poiré, or tea are interesting because they tend to tune down the richness of the cheese, while keeping their own identity at the same time.

LEARNING TO LIKE

We can—and occasionally even must—learn to appreciate harmony. For instance, if signals are intricately interwoven and hard to recognize, our brain may initially fail to see the harmony. However, with some help and training, we can ultimately learn to recognize the beauty. Give this some extra thought. Is this not the ultimate challenge for hospitality? The perfect sommelier or host gives the extra information that helps the brain receive the information in such a way that fluency is increased.

This role of hospitality is underestimated. It is the essential difference between service and hospitality. Service is all about transporting plates, filling glasses, and answering questions. Hospitality is about adding value to all of this by giving information that helps the guest understand what is behind what is served and to convey the passion of the chefs and winemakers. As such, hospitality is a part of flavor and can be considered as one of the extrinsic factors. Its role should certainly not be underrated, as we have discussed in Chapter 3.

Learning about and accepting flavors are also seen with the so-called acquired tastes. A liking for beer, Brussels sprouts, coffee, and plain chocolate takes time to develop. Wine tasting is also often mentioned in this respect. People can learn to recognize flavors and build up experience. In the process, it is likely that preferences and liking are going to shift. Elements that experts consider as positive may be negative drivers of liking for the average consumer; on the other hand, amateurs may like flavors that experts consider defective.[12,13]

CULTURAL CONTEXT

Processing information is one side of the story; interpreting is another. In a wine and food tasting, it is quite possible—even likely—that most people will share the same experience, but that does not mean that they have the same opinion about the combination. Suppose, for instance, that a certain dish is on the coating side, and it is served with a wine that is more contracting. The expected effect is that the acidity of the wine is enhanced; it will seem more contracting than before. Now, even though this reaction is experienced by most, opinions about the change may not be the same. Presumably, most people will be rather critical, yet others may appreciate the wine turning more contracting than it was before.

Knowing this, you should realize that the semidry wines such as the German halbtrocken or feinherb types have to be watched carefully. They are two-sided, which basically implies that if the accompanying dish is on the acidic side, the sweet side of the wine will clearly show. Similarly, when the dish is sweeter, the acidity of the wine will show more. If such a situation occurs, it is quite likely that people may be confused. Some will be enthusiastic, others may disapprove, and some others will find it fascinating that this phenomenon happens. Fully in line with the theory of harmony, sweet and sour dishes combine best with semidry wines. Indeed, certain types of Asian dishes go well with these wines.

Note that we say "certain types of Asian dishes." Asia is a huge continent with many differences in climate and culture. This diversity is reflected in their kitchens and in their flavor profiles. Some dishes are more on the sweet side, and others can be very spicy; there is quite a variety in flavor richness. One general trait is that umami components are mostly used as a salt. Cooking is an expression of a culture. Therefore, as beverages and foods interact, it is important to know what the dish wants to express. A wine with residual sugar can have an unwanted effect on the flavor palate of a dish. It helps to understand the cultural context of dishes or the philosophy of a chef.

From a cultural point of view, Asian and European cuisines are not only different in flavors. A fundamental distinction is that in Europe the plate is central, and in Asia it is the table. Many Asian cuisines make use of side dishes with condiments such as soy sauce, ginger, or wasabi. It is up to the guest to decide whether to use them and how much. In the European tradition, the chef decides, and sometimes there is even discussion whether salt should be on the table. Clearly, from a gastronomy point of view, it helps when the flavor profile is stable; this is not the way people eat in Asia, however.

COOKING TOWARD WINE AND PRECISION

There is an interesting difference between chefs and pastry chefs. Pastry chefs always use scales, which are used to weigh ingredients to achieve the desired results. It is hard to taste bread dough or cake batter or a mixture for making ice cream. As a pastry chef, you learn to be precise. Regular kitchen chefs have a different approach. They are usually trained to assess and adjust the final product. Their "pinch of salt" or "reduce the sauce to the right thickness" says it all.

This is not about passing judgment on either approach. What needs to be stressed is that small changes can have significant effects on the flavor profile. This makes it possible to cook *toward* a wine. This is important to realize. At home, you can easily open the bottle that is going to be drunk, have a sip, and then make the final adjustments to the dish. You can choose herbs or spices in line with the profile of the wine. Adjust mouthfeel with the sauce. Determine that "right thickness" as a function of the wine. Reduce some more, or add some cream if coating is required. Or do the opposite—stop reducing, skip the cream, and add a touch of acidity to move the dish to the contracting side.

In professional situations, this last-minute adaptation is hardly practical. It would be quite a challenge to let guests all make their individual choices and then to start cooking. The more practical method is to ensure that sommeliers and other people in service are well aware of the flavor profile of the dishes that are on the menu. Preferably, they will have all tasted these dishes, enabling them to give sound advice to the guests. There is also a more fundamental way of organizing gastronomy: develop dishes together with the wine and put them both on the menu. In the ideal situation, these wines are offered by the glass. It is the best of both worlds: guests make the choice of the dishes they like and do not have to compromise by choosing one wine that must accompany all the chosen dishes.

The gastronomic development of dishes paired with specific wines can be a fascinating experience; ideally, it is done with a group of people from the kitchen and from service. In most cases, the chef will propose the dish. It is tasted and potential wines are tasted with it. This may lead to adaptations of the dish—changing a garnish, leaving something out, adding something else. It is crucial to have the kitchen involved because its awareness of how dishes can be influenced is increased. This also requires chefs who are willing to change recipes in favor of their combination with a certain drink. Ideally, they know about the principle of gastronomy and realize that the good match of food and beverage enhances the guest experience.

Now suppose there are some cases of a wine that are close to being past their best. The gastronomic thing to do is take that wine and compose a dish for it—ideally, one that is truly good for the wine and sells it. It helps to be proactive and not wait for the problem to arise. It is quite feasible to put a wine—or another beverage—first and design a dish to go with it. In that case, you start with properly defining the flavor profile of the beverage and go from there. This is an interesting culinary experience because it stimulates creativity and thinking "outside the box." Many people—chefs included—are full of automatism when cooking. These are forms of learned behavior that a person develops through the years to reduce the burden of constantly having to make choices. When the mission is to design a dish to go with a specific beverage, it helps to be free of such automatisms. A useful approach is to taste every single ingredient that you would want to use in combination with the beverage. Everything that works can potentially be used; ingredients that have negative effects should be excluded. This is also a good system to find "problem makers" in an existing combination. Every once in a while you come across something that does not quite fit. It is not directly identifiable, yet it is undesired. In order to find it, try the different elements (main ingredient, sauce, herbs, condiments) of the dish together with the wine. If the problem maker is found, it can be changed.

Once a combination has been developed, the specifics of the dish and the beverage are registered. This goes beyond the recipe and should include the amount of sauce and garnishes. Preferably, a picture is included to show what the dish looks like. For the beverage, the serving temperature and the glass are noted and, of course, which year it is. Wines made traditionally tend to be somewhat different from harvest to harvest.

EXPENSIVE WINES

For many people, premium wines are at the summit of gastronomy, and many of these wines find their way to great restaurants or beautiful cellars. Mostly, they will be consumed with food. Now, I have heard people claiming that great wines will always match great food. Unfortunately, this is not the case; foods can easily have a negative impact on premium wines. Arguably the effects are even more undesirable: you do not want anything to interfere with the pleasure of enjoying a great wine.

The situation worsens in the case of mature classical wines. They are fine with "classical" dishes—without added sugars, excessive creaminess, exotic herbs and spices, just to name a few. In fact, they are perfect with dishes that are rather pure and simple. A great classic red will always show well with a good piece of meat, some au jus, and mushrooms. Serving great wines with good but somewhat simple dishes is like giving them center stage and creating a situation for them to shine.

There is a nice philosophical rationale for doing this: if a premium wine is served, why put it in competition and risk that it will lose the battle? For the same reason, it is debatable whether two great wines should be served side to side with a dish. For whatever reason, one is likely to be found better than the other; if one wine is served by itself, there would be no discussion.

Just the same, premium wines are not for every occasion. It is similar to cars. For some people, possessing and driving a luxury sports car is fun. However, these can be rather awkward on a bumpy road and are ill suited for many purposes. This is called "fitness for use" and is a concept not often considered in gastronomy. The gastronomic version would be: a cheap wine that fits the dish and the circumstances gives more pleasure and satisfaction than an expensive one that does not fit.

MARRIAGES MADE IN HEAVEN

The *Oxford Dictionary* considers the combination of strawberries and cream so successful that it is mentioned as an example of a marriage made in heaven. In gastronomy, some wines have also been labeled as perfect partners with specific foods. Lists will feature oysters and Chablis, lamb with Pauillac, Roquefort with Sauternes, and so forth. After all that has been said, it is hard to accept such generalizations. Any marriage requires at least some kind of an effort to make it work. Even though it is true that there are indeed wines from Chablis, Pauillac, or Sauternes that can be perfect with "their" food, it is a challenge to find the right one. Within each region, you will find lighter and heavier, younger and older, simple and breathtaking wines.

On the food side, most Roqueforts are quite comparable in flavor; Sauternes will match provided it is rich enough. The variety in oysters is already too big to be safe with Chablis. Farming methods and size have a great impact on flavor. For a Chablis to match, oysters need to be rather fleshy and not too salty. Furthermore, refrain from the popular add-ons such as lemon, raw shallots, vinegar, and Tabasco because they are likely to kill the wine. If you do all that, have a regular Chablis from a good year; it will be better than a premier or a grand cru. Even more is required to make the marriage of lamb and Pauillac work. We need to prepare the lamb, and it is mostly served with a sauce. The flavor profile is going to be influenced by what has been done to the meat and by the sauce and the other elements on the plate. The chance that a Pauillac is the ideal partner is just as likely as a red wine from Greece, Argentina, or China. Make a curry of it and a German halbtrocken Riesling or a rich Gewurztraminer from Alsace is the better choice.

Which makes you wonder, how did Pauillac become the preferred and much acclaimed partner? The most probable explanation is that there is actually *agneau de Pauillac*, lamb from Pauillac. It is an AOP, a protected product. However, if you go to Pauillac itself, a town with some 5,000 inhabitants on the border of the river Gironde, you will not find any meadows with sheep, but vineyards. Pauillac is arguably the most esteemed of the wine appellations of Bordeaux, and the land is therefore far too precious to raise sheep. In fact, the AOP rules for the *Agneau de Pauillac* indicate that they can be produced all over the department Gironde, which happens to be the largest of France (10,000 km²). Therefore, as far as the marriage is concerned, any of the other red wine regions of Bordeaux would also have qualified.

REGIONAL WINES WITH REGIONAL FOODS AND OTHER CUSTOMS

It seems evident that regional foods and wines should have a natural fit. Nevertheless, it is not something to take for granted. The main reason is that things change over time. Wines, especially, have changed. Modern insights, yeasts, and techniques are applied everywhere and have profoundly changed the profile of many wines, independent of their appellation and grape variety. Consequently, many of the regional wines do not taste like they used to anymore.

One of the most prominent regions where this has clearly happened is Burgundy. There used to be a time—and it is even not too long ago—when it was completely acceptable to add wines from the south of France, or even Italy or Algeria, to the wines of Burgundy. Even now, Burgundy wines have an image of being "heavy." However, this is not the character of the Pinot noir. Rather the opposite; fragrant and elegant would be more fitting as general characteristics for the modern red wines of Burgundy.

The same can and must be said about the regional dishes. They have been "improved" as well. Today, we expect chefs to add personality to their dishes. As we have seen, small changes can have a big effect; a match with a regional wine cannot be taken for granted from the food side either. Hence, an ancient Burgundy might have been a good choice with a traditional *Boeuf Bourguignonne*, but these days you would have trouble finding one.

In this respect, it is important also to consider culinary automatics: habits that we have grown accustomed to and always repeat, without really knowing why. In Holland, it is customary to have raw onions with the famous herring. Likewise, smoked salmon is often served with lemon or horseradish in some countries. There are many other regional and cultural gastronomic habits. In most cases, the garnish has—or had—a function; a reason for having it. In the case of the herring, this stems from the time that cooling was not very possible. The fats of the herring or the salmon oxidized rather quickly, which gave a specific strong fishy flavor that was efficiently masked by the raw onions. Now that everything can be kept cool, the technical problem has been solved. Nevertheless, the culinary custom has remained.

ROBUST AND ELEGANT

Regional cuisines and dishes share a characteristic; they are what can be called "robust." This is what sets them apart from what may be called "gourmet food." The opposite of robust would be "'elegant," which is indeed a word that can be used to characterize the dishes of the great restaurants of the world. How do these concepts relate to gastronomy? An obvious logic is that refined, elegant dishes call for refined, elegant wines. Likewise, there is nothing wrong with having a good but plain wine with a good but ordinary dish. This becomes more complicated when the question is raised as to exactly what makes a wine or a dish elegant or robust.

In an effort to resolve this issue, I once wrote a letter to the world's most renowned winemakers and journalists. To my surprise, there was no clear-cut answer. Many of the respondents associated elegance with refinement and complexity, as if it were the ultimate goal of the wine-maker. Needless to say, for this group, robust was synonymous with boorish, coarse, and uninteresting. Others had more appreciation for robust wines and wondered whether the two concepts were each other's opposite. For this group, the issue was not quality but a question of style. Elegance implies playful, subtle, delicate, refined, aristocratic, light, and distinguished. Robust is characterized by words such as rustic, heavy, massive, earthy, pure, strong, rich, full, straightforward, and outspoken. The summit of elegance is expressed by the word "finesse"; with robust, that word is "opulence." Indeed, they are not opposite but represent different worlds.

Compare this to furniture. What makes a table or a chair elegant or robust? The wood that is used in elegant furniture is thinner, and it has probably a finer grain; oak is more robust than mahogany. Both can be strong and valuable, but they just do not belong together. An elegant chair looks funny next to a robust table. The environment is also important. The robust oak table might be beautiful in surroundings where it fits but be terribly out of place somewhere else. With robustness, the basic character remains clearly recognizable. This is also seen in clothes. Compare a woolen dress to one of silk. In silk, the threads are so thin and intertwined that the individual fibers are no longer visible. The woolen knit, on the other hand, clearly shows the yarn. The dress can be beautifully made out of the best wool, but its structure is coarser than the silk garment. And how about wearing a silk dress in a rustic environment?

Before taking this back to the world of gastronomy, another phenomenon must be mentioned: buffer capacity. A beautiful country table with dents and scrapes remains a beautiful table. We consider that these dents belong there. An elegant table that has a scratch loses a part of its beauty. Likewise, we immediately notice if a ballet dancer is one pound overweight. In other people, a few pounds are hardly noticeable. Apparently, robust products are less easily out of balance; elegance is more specific and demanding.

In terms of the flavor profile, elegance is mostly more on the contracting side but not dry; flavor intensity is seldom very high and there will always be fresh flavor tones. For robustness, mouthfeel is rather coating or dry, flavor intensity can be high, and ripe flavor tones are more likely to prevail. Furthermore, robust flavors are recognizable. This usually has to do with the basic materials used. It is hard to make an elegant dish with hare, eel, or mackerel. This is much easier with lamb, turbot, and sole. In cooking, techniques such as poaching, steaming, and pan-frying belong to elegance. Baking, grilling, and braising give a more robust, dominant flavor with a bigger buffer capacity. This does not mean that these last techniques cannot give elegant results. It is all in the hands of chefs and winemakers, who are capable of intertwining flavors such as in silk. Subtlety is the word. No flavor is dominant; together they form a new identity.

Subtlety requires the taster to make an effort and pay attention to the details; otherwise, he is likely to overlook some of the flavors that are there. With the robust and rustic flavors, this is not necessary: they come to you. This implies that the choice of elegant or robust flavors all comes down to the one who tastes. To really appreciate elegance, take your time. If you do not have the time or your mind is somewhere else: go for robust.

BEER AND FOOD

Although wine is the dominant choice with foods, let us not forget beer in the combination. There are combinations where matching wines are hard to find. Certain cheeses have already been mentioned in this respect. Additionally, because many wines have an alcoholic content of more than 13% these days, there is a place for low alcohol alternatives at a dinner where several wines are drunk with the dishes. This is good news for the use of beer in gastronomy. Evidently, the match of beer and food demands the same attention as selecting a wine.

Furthermore, when we discuss the use of beer at a restaurant table, it is the specialty beers that we are interested in. The small and sometimes local breweries deserve our attention and support just as with wine. Beer and wine are not the same; their differences lead to the following recommendations:

- Watch the type of bitter. The bitterness of beer comes from hops, a type vegetal bitter that we do not find in wines. In food, however, we do find similar bitters in members of the cabbage and lettuce families. This implies that beers may come to the rescue when wines run into trouble. Beers and dishes with prominent vegetal bitters find each other. The bitters of beer can also be compensated with fat or proteins. In general, fish and beer go fairly well together. Sweetness in food is not as effective in compensating for bitterness.
- Watch the flavor richness. Light beers are suited for dishes that are low in flavor richness, such as salads and simple fish dishes. Such dishes are dominated by many wines, and a light beer is a useful alternative. Salt seems to work well with beers.
- Watch the color of the beer. The darker beer types seem easier to combine with foods that are characterized by the Maillard reaction. This stands to reason when you realize that the same effect is also responsible for the brown color of the malt that is used in making these beers.
- Make use of the bubbles. In line with what has been said about the CO_2 in wine, the carbonic acid is effective for cleansing the mouth and breaking through the coating mouthfeel (such as from a buttery sauce or a poached egg). The bubbles also help certain beers in combination with coating cheeses.
- Watch the acidity. The pH of most beers is around 4.1–4.5, while many wines are between 3.1 and 3.5. As acidity in beverage and food adds up, it is easier to find a beer to match a dish that has high acidity. On the other hand, beers with high acidity (Geuze or Weizen) are not easy to combine with dishes. The combination with the CO_2 makes them react stronger than wines, which makes finding the right balance even more of a challenge.

- Watch the sweetness. Sweetness in dishes is often a problem because most beers are not sweet enough to match. Only moderate levels of sweetness in dishes are possible provided that the beer has about the same level of sweetness, just as with wine.
- Beware of spicy food. The pilsner/lager types of beer tend to intensify the spiciness of Asian food. The better choice would be a sweeter beer or a dark type.

SUMMARY

The world of flavor has seen many developments and calls for new guidelines for combining food and beverages. The following recommendations have been developed based on the universal flavor factors:

- Mouthfeel needs to match: that is, contracting fits contracting, coating fits coating
- Flavor intensity needs to match
- Flavor style needs to match: fresh fits fresh, ripe fits ripe

The following may be added:

- Contracting before coating
- Fresh before ripe
- Low intensity before high intensity

The challenge of finding good matches is in recognizing and interpreting the flavor factors. It is imperative to taste. When tasting, pay particular attention to the dominant factor—the flavor factors that are the most important. The dominant factors of the food and the beverage need to match.

The basic flavors, including spicy, have a big influence on wines. Acidity and saltiness are handled quite well, but the amounts need to be observed. Sweetness is a problem for wines that have no sweetness of their own. They are likely to turn sour. In the case of sweet dishes, the most fitting wine is preferably a bit sweeter than the dish. Spiciness is different from the rest. In fact, some wines react fairly well—especially those wines that are higher in alcohol (>13%).

In mouthfeel, the role of fats, especially in sauces, needs to be observed. Fats and wines do not match easily. However, if coating originates from starches, finding a match with wines is easier. Tannins have a drying effect and fit well with dry foods. CO_2 is a useful component because it can cope with extreme coating. This is shown, for instance, with some cheeses. Cheeses in general are rather particular in gastronomy. White wines perform better primarily with the ones that have creamy and soft textures. The suggestion to match with a local wine only helps in some cases. The same can be said for the combination of regional dishes and regional wines.

Finding harmony is essential in gastronomy. The explanation is found in the way our brain is organized. A prerequisite for liking is that sensory information can be processed quickly. Something to monitor is that harmony can become boring. If used functionally, a contrast can prevent that. There is an important role for the people in service. By explaining the flavors and the reasons for their combination, these people can truly enhance the guest experience.

NOTES

1. Rosengarten, D., Wesson, J. (1989). *Red wine with fish: the new art of matching wine with food.* New York: Simon & Schuster, p. 11.
2. http://www.cellartracker.com/clASSIC/wine.asp?iWine=680492
3. http://www.chateauclassic.com/produit.asp?id=390725
4. Obviously, it would also help if the writers of tasting notes gave systematic and adequate descriptions.

5. Chandrashekar, J., et al. (2009). *The taste of carbonation.* Science (326):443–5.

6. McRae, J., Kennedy, J. (2011). *Wine and grape tannin interactions with salivary proteins and their impact on astringency: a review of current research.* Molecules (16):2348–64.

7. Peyrot des Gachons, C., et al. (2012). *Opponency of astringent and fat sensations.* Current Biology 22(19):R829–30.

8. Barasch, M. (2000). *Theories of art—from impressionism to Kandinsky.* New York: Routledge, pp. 332–3.

9. Bavaresco, N. (2004). *Harmonic compositions of complementary colors according to their lightness degree.* AIC 2004 Color and Paints, Interim Meeting of the International Color Association, Proceedings 235.

10. Rameau, J.-P. (1722). *Traité de l'harmonie, réduite à ses principes naturels.* Paris: Broude Brothers.

11. Ishizu, T., Zeki, S. (2011). *Toward a brain-based theory of beauty.* PLoS ONE 6(7):e21852. doi:10.1371/journal.pone.0021852.

12. Hughson, A. L., Boakes, R. A. (2002). *The knowing nose: the role of knowledge in wine expertise.* Food Quality and Preference (13):463–72.

13. Delgado, C., Guinard, J.-X. (2011). *How do consumer hedonic ratings for extra virgin olive oil relate to quality ratings by experts and descriptive analysis ratings?* Food Quality and Preference (22):213–25.

CHAPTER 7

Food Appreciation
and Liking

INTRODUCTION

So here we are. Gastronomy has been introduced as the science of flavor and tasting. We have seen how tasting works and have made clear that all our senses are involved in the perception of flavor. Flavor has been defined as a product quality, which consequently makes it a domain of the natural sciences. We have elaborated on how flavor can be described and can be looked at from all angles in foods and beverages. We have philosophized about flavor matching and have shared our experience from the last 20 years.

One last challenge lies ahead. The study of gastronomy boils down to the crucial question: "Do we like what we are eating or drinking?" Liking, just as disliking, is a crucial driving force in the selection of the foods and beverages that we consume. Pleasure is the incentive for making us continue to eat or to drink, and it ultimately forms our behavior. Similarly, on the negative side, disgust is also a strong force. Some of the answers to liking and disliking are found at the objective level; not everything fits together, and there are proper methods and sensible guidelines that ensure good results. This is the part of flavor that can be learned—at least to some extent—and, hence, we have devoted the previous chapters to it.

For other answers on liking, we need to look at the human—the individual—subjective side. What happens in the process of the human perception of flavor? Our senses register the flavor, but the information is processed and interpreted in the brain. Therefore, liking beauty and appreciation are functions of the brain, not of one of the senses. The senses are subsidiary to the experience. In an effort to comprehend concepts such as liking and palatability, we need to move from the receptor level to the brain. This requires modesty, humility even. After all, the human brain is a huge machine with billions of neurons interacting via electrical and biochemical signals. Therefore, it is inconceivable to grasp the complexity of the human brain in just a few pages of a book.

Nevertheless, however daunting, from the perspective of gastronomy it is fascinating to explore how our brain functions and to try to make a start in making inferences about the way people desire, dislike, perceive, and enjoy foods and beverages. Especially when the added value of foods and drinks is high, as happens to be the case in hospitality (hotels, restaurants, bars), it is essential to make food *desirable*. This is an important key to commercial success. Consequently, it is important in hospitality, and thus in gastronomy, to know more about the pleasure side of food. This chapter provides insights that elucidate parts of the story. It covers subjects such as wanting and liking, palatability and liking, food selection and harmony, and introduces new concepts such as the aesthetics of gastronomy and "Guestronomy." These new insights may inspire future research in gastronomy.

TASTING AND THE BRAIN

FROM RECEPTOR TO THE BRAIN

The brain is the place where all sensory information comes together; the processing takes place, conclusions are drawn, and actions are initiated. By nature, the flavor system is extremely complex. All major sensory systems—gustatory, smell, touch, vision, and hearing—are involved. The first three deliver information on the intrinsic qualities of the food and beverage. Vision and hearing may have a big impact on the registration of the intrinsic elements and are also involved in the registration of the extrinsic qualities. Both the intrinsic and the extrinsic clues are important. The systems are intertwined, which enhances the complexity.

The receptor neurons in each sensory system may have different functions and operate differently, but they all do the same job: converting a stimulus from the environment into an electrochemical nerve impulse, which is the common language of the brain. Basically, sensory systems such as vision, audition, olfaction, and gustation share many common features.[1] From our review (Chapter 2) of the neural systems that are involved in the registration of flavor, one can draw the conclusion that there is sophisticated knowledge about the way components of flavor are registered orally and nasally, although not all aspects have been uncovered as of yet. Nevertheless, the receptors are only the first level of registration (see Figure 7.1). To quote Spector: "With so much recent focus on the transduction in the gustatory system, it is easy to be seduced into thinking that the solutions to the neural basis of gustatory function are to be found at the level of the receptor cell. But this level represents only the first stage of signal processing, albeit an absolutely essential one."[2]

Following the neural pathways that ultimately lead to the recognition, interpretation, and evaluation of flavor begins with a tastant that interacts either with proteins on the surfaces of cells or directly with pore-like proteins called ion channels. These interactions cause electrical changes in the receptor cells that trigger them and lead to chemical signals that ultimately result in brain impulses.[3] Different chemicals use different channels. Na^+ ions (from salt and acidity) are able to enter taste cells directly, resulting in the depolarization of the cell, without the interference of second messengers (the taste stimulus is the first messenger). Besides chemicals that produce salty and sour flavors, pain-like sensations registered through nociceptors of the trigeminal system also act directly through ion channels.

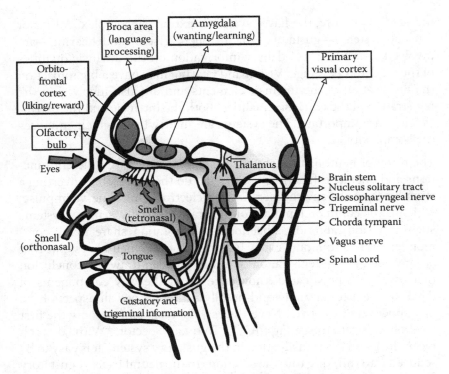

FIGURE 7.1   NEURAL PATHWAYS OF FLAVOR TO AREAS IN THE BRAIN.

Sweet and most bitter flavors, umami, many odorants, as well as fat are registered on the outside of cells. They trigger changes within the cell by binding to receptors on a gustatory cell's surface that are coupled to cytoplasmatic molecules called G-proteins. These generate intracellular compounds called second messengers. Second messengers alter electrical properties of the gustatory cell and modulate the release of neurotransmitters. The G-protein of the gustatory system was called gustducin because of its similarity to transducin, the protein in retinal cells that transduces the signal of light into the electrical impulse that constitutes vision.[4]

Flavor registration pathways may thus be classified as two kinds: the ion-channel group (IC) and the G-protein-coupled receptors (GPCR). Within each group, different kinds of gustatory receptor cells have been identified with specific qualities. Research in this field is rapidly progressing. The gustducin-linked receptors have now been identified as a family of some 25 GPCRs. Some are more present in fungiform papillae, others in foliate or circumvallate papillae. The olfactory systems work only with GPCRs.[5]

However and whatever, the flavor information is transduced into electrical signals and sent to the brain. It first reaches the reception area that every system has: the nucleus of the solitary tract (gustatory system), the olfactory bulb (olfactory system), or the thalamus ventral posterolateral nucleus (trigeminal system), and those of vision and hearing. Because these primary cortices are still a part of their own neural pathway, it is not flavor that is being identified. For a full identification of flavor, the systems have to converge. The diagram developed by Rolls (Figure 7.2) shows the individual pathways and the brain areas where they come together.[6]

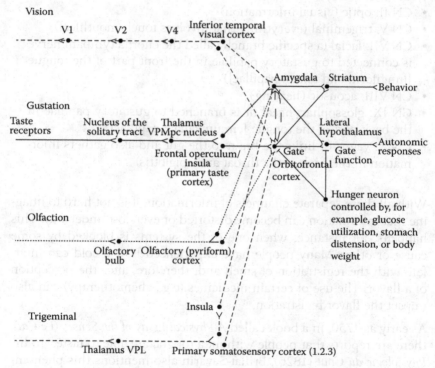

FIGURE 7.2 SCHEMATIC DIAGRAM SHOWING SOME OF THE GUSTATORY, OLFACTORY, TRIGEMINAL, AND VISUAL PATHWAYS TO THE AMYGDALA AND THE ORBITOFRONTAL CORTEX.

SYNTHESIZING NEURAL INFORMATION

The human body is in large part operated by the brain. This uses up a lot of energy; the brain constitutes only 2% of the body mass, but it consumes 20%–25% of the daily calories, even when the body is asleep. The brain gets its information and impulses to act from the sensory nervous system. A distinction is made between cranial nerves (CN) and spinal nerves (SN). The cranial nerves emerge directly from the brain, and seven of the twelve pairs are involved in the transduction of sensory information, while others are designed to operate muscles. Some nerve systems do both. The cranial nerves that are relevant for flavor perception are:

- CN I: olfactory (orthonasal and retronasal odorants)
- CN II: optic (visual information)
- CN V: trigeminal (everything that involves touch/mouthfeel)
- CN VII: facial (a specific branch called the chorda tympani nerve is connected to gustatory papillae in the front part of the tongue [fungiform and foliate papillae])
- CN VIII: acoustic (hearing)
- CN IX: glossopharyngeal (it is branched to gustatory papillae in the back part of the tongue [circumvallate papillae])
- CN X: vagus (it links the brain to the gut and also gathers information in the back of the throat and epiglottis)

With all these separate channels of information, it is not hard to imagine that information can become distorted or even lost under way. This happens, for instance, when one of the systems is blocked by some cause or other. Many people have experienced how a cold can interfere with the registration of smell and, therefore, alter the perception of a flavor. The use of certain medicines (e.g., chemotherapy) can also impact the flavor registration.

As early as 1750, in a book called *A Physical Essay of the Senses* (*Le Cat*), there are reports that people without a tongue are able to taste.[7] In his *Physiologie du Goût* (1826), Brillat-Savarin also mentions this phenomenon. This is not surprising considering the nerve systems that are all involved in flavor perception in some way or other. In fact, there appear to be many examples of individuals with extensive nerve damage to the tongue whose flavor registration remains more or less intact. For example, a branch of the facial nerve, the chorda tympani, can easily get damaged (e.g., during ear surgery) because this nerve crosses the medial surface of the eardrum. When this happened, in some cases, individuals did not experience a decrease in tasting abilities, although the taste function did not return to the area that was serviced by the damaged nerve.[8]

The explanation is that the chorda tympani is not the only cranial nerve transducing gustatory information. The back part of the tongue, about one-third of the surface but representing 50% of the taste buds, is mainly linked to the glossopharyngeal nerve. Apparently, the human nervous system has the capacity to bypass barriers by activating other routes and thus keeping flavor registration more or less intact. Flavor perception appears to result from a system with a great deal of redundancy, such that even relatively extensive losses often go unnoticed.[9]

However important the issue of transduction of flavor stimuli may be, and despite the scientific attention it has always received, it is doubtful that the ultimate answers on the functioning of the human registration of flavor are to be found there. Furthermore, until recently, sensory perception (of all senses) has been studied as isolated modalities. Thanks to recent developments in research using modern functional magnetic resonance imaging (fMRI) methods, the question has been raised as to whether perception in general must be seen as a system of separate modular functions, in which the different senses operate independently of each other. Many daily events are registered simultaneously by several senses at the same time, for example, by the eyes and the ears. Think of how the experience of viewing a motion picture can be enhanced by music and "third" dimensions. In human beings, the language capacity is developed to a uniquely high level, enabling man to reflect upon everything that is perceived in the brain. This implies that in the brain different types of sensory information are not only integrated but are also "labeled" with words (at least partly) to enable communication. This is the essential difference between flavor and flavor perception.

CROSS-MODAL INTERACTION

In normal eating situations, flavor perception is by definition a three-dimensional experience. We taste with our eyes, ears, tongue, nose, and touch; these pathways deliver parts of the information, and they interact. The word to use is cross-modal. A study on quality evaluation of crisp bread showed that subjects needed olfactory, tactile, and auditory information to recognize samples.[10] Such cross-modal interactions among the senses are likely to be the rule and not the exception in perception. As soon as more sensorial elements get involved, new dimensions open up and perception is likely to be altered, or perhaps we should we say "completed." Scientifically, it would be helpful if humans were neutral registrars of these components, but clearly this is not the case. Humans are not organized to function like a sensory laboratory.

Cross-modal interaction of the senses with regard to flavor has already been mentioned in the section "Carry-Over Effects" in Chapter 2, but it is worth further elaboration. Because the flavor system relies upon an active involvement of different senses, cross-modal effects are abundant. Sensory interactions occur each time we eat or drink. Gustatory flavors interact with each other. Flavors can be enhanced or suppressed. Salt and umami are known as flavor enhancers, and that is exactly what they do: all kinds of flavor tones are enhanced by salt or umami. To a certain extent, acidity does this as well. In contrast, fat and sweet suppress, for instance, bitterness. Other sensorial elements of foods and drinks such as the texture or smell are also known for their cross-modality. Changing textures is a main attraction of molecular cuisine, and aromas can modify the perception of acidity, saltiness, or sweetness. Smells that match the "right" color are perceived as being more intense than in combination with another color.[11] Even if we limit ourselves to the interaction of the intrinsic components of flavor, there are numerous examples of cross-modality.

In the section "Intrinsic and Extrinsic Components of Flavor" in Chapter 2, the extrinsic components of flavor were introduced. Examples were mentioned that involve, for instance, vision and hearing. In Dubourdieu's study, the white wines that were colored red were described differently by subjects.[12] Another well-known example is to change the perception of the sweetness of yogurt by coloring it red. The same applies when this same red yogurt is served in a black bowl. The contrast makes the red look redder and the taste sweeter. In general, presentation is important. Indeed, the plate can have an influence on the flavor of a dish. Food served on a star-shaped plate is perceived as bitterer than when served on a round plate. Yogurt was judged to be more dense and expensive when it was eaten from a heavier bowl. These heavier bowls even made people feel fuller. Such strong relations have important implications for food design and packaging.[13,14,15] Aesthetic principles also rule in the gastronomy world.

Cross-modal interaction goes further. The environment influences perception; ambient lighting is a good example. In a tasting of Riesling wines in a winery in Germany, the color of the room was changed from white to blue, red, or green while 200 wine buyers rated a certain wine. The wine itself was served in a black glass to prevent the influences of the color of the room upon the perceived color of the wine. Under blue and red lighting, the quality of the wine was rated significantly higher than under green or white lighting. People were even willing to pay 50% more for the wine tasted under red lighting compared to the same wine under green or white lighting. Further research revealed that blue and green room lighting made wines taste spicier and fruitier, while the red lighting made the wine taste nearly 50% sweeter.[16] Good-looking food will induce one to eat it, just as seeing people enjoy a certain food will do so. Just plain seeing foods and having foods on hand leads to an increased intake.[17]

Other experiments show the influence of environmental sounds. Music in restaurants, for instance, may well prove to influence the flavor experience. People enjoyed eating oysters much more when hearing breaking waves than farmyard sounds. Conversely, a dish of bacon and eggs tasted more "bacony" when listening to the sound of sizzling bacon than to a farmyard of clucking chickens. The earlier-mentioned dish of Heston Blumenthal, "the sound of the sea" (see the section "Very Cold Cooking: Liquid Nitrogen and Other Molecular Novelties" in Chapter 4), is a good example of how this knowledge can be set to use.[18]

Unadulterated flavor perception can be expected in an environment where there is little distortion. White booths with infrared lighting are ideal for conducting scientific tasting experiments—even though it is questionable whether results that are reported based on such an environment reflect a real-life situation. And even then, tasting in a scientific setting will not be completely neutral. Humans are involved because of their background, habits, and previous experiences. It is hard to block the impact on flavor perception of negative or positive experiences from the past or from expectation. If we ever think that we have turned ill after eating or drinking something, we are likely to be careful the next time. Looks can also be deceiving, and information can be labeled by all kinds of beliefs. We need proteins, but many Western people object when these proteins are a part of worm, maggot, dog, or a snake; Indian people revolt at the idea of eating a sacred cow, and Anglo-Saxon people feel so about their "sacred" horse. This demonstrates that even aspects that have little to do with actual flavor may indeed influence liking. The role of our brain should not be neglected. In eating and/or drinking, the brain synthesizes all the sensory information it gets and connects it to previous experiences and to expectancy.

FOOD LIKING AND PALATABILITY

Looking more closely at the brain and its pathways in Figure 7.2, you see that information is processed in two centers, the amygdala and the orbitofrontal cortex. Their functions are quite different, and this has to do with the reasons why we eat: eating as a biological necessity versus pleasure eating. The first motive is controlled by the amygdala. The orbitofrontal cortex is involved in the pleasure part of eating.

WE LIKE WHAT WE NEED

The first and principal reason for eating is a biological one: food is fuel and building material, which are needed to keep our system going and intact, in terms of energy and quality. It is no coincidence that in many languages there are sayings such as "hunger makes the best spice." The French equivalent would be *à qui a faim, tout est pain* (for someone who is hungry, everything is bread). Likewise in Spanish they say *a buen hambre no hay pan duro* (if you are hungry there is no old bread). In Dutch, it would be *honger maakt rauwe bonen zoet* (hunger makes raw beans taste sweet). Indeed people do not worry about hedonic aspects such as palatability when they are truly hungry or thirsty. The functioning of the biological system is potentially in danger; and apparently our system does not want something as trivial as palatability to stand in the way.

If we take this one step further, it is quite reasonable to assume that from an evolutionary, biological point of view, liking is the way to attract people to the nutrients that they need: salt, sweet, fat (energy), and umami (proteins and their amino acids), or to prevent us from eating things that may potentially harm us: disliking bitter (poisonous alkaloids). In the case of salt, this system appears to be particularly sophisticated. We need salt but not too much. We like it at levels that are beneficial to the body, but this changes to a powerful adverse reaction when the level gets too high by activating cells that provide the sour and bitter taste. This appetitive/adversive balance helps to maintain appropriate salt consumption.[19]

Furthermore, the role of vision and smell is important in the primary function of liking and disliking: we need our eyes to pick the right fruits, and our nose to smell ripeness or spoilage. These are all fundamental functions, largely developed through evolution. It is a system that works perfectly and has kept us healthy. And it is strongly related to the brain: we know what we need and have a strong sense of survival. Ultimately, we eat anything to survive, even fellow humans.

Consequently, the human body may continuously be monitoring the need for certain nutrients. A fascinating illustration is pica: children who are deficient in a certain mineral begin eating dirt. Apparently their mind knows what the body needs and solves the problem by making them like what they would normally never eat. The peculiar food habits some women experience during their pregnancy may also be explained on this basis. It may well be that there are many physiological processes going on without ever being actively and consciously noticed. This implies that subconscious physiological pathways may drive food choices, and that eating and drinking as a motivational state may coexist with physiological mechanisms. It also underscores the importance of proper food identification and recognition by consumers because they possibly may—subconsciously—want specific nutrients to replenish physiological needs.[20]

Quite another influence on eating may be related to the sympathetic nervous system (SNS), which modulates changes, such as in heartbeat and blood pressure. It is hypothesized that the SNS also plays an important role in food intake. A low sympathetic activity, which is observed in fasting and in low temperatures, is associated with a low threshold for eating and, when food is available, more food is eaten.[21]

If appetite is defined as a desire for food, and satiety as the fulfillment of this desire, it is reasonable to assume that both are part of a simple regulative system. Several peptides such as neuropeptide Y (NPY), orexin-A, orexin-B, β-endorphin, and galanin are reported to stimulate food intake. NPY most strongly induces the motivation to eat. Peptides such as leptin, corticotropin-releasing hormone (CRH), cholecystokinin (CCK), enterostatin, and bombesin are reported to decrease food intake. Of these, leptin is often mentioned as an important peptide.[22] The role of peptides such as NPY and leptin is very much related to energy needs, and they are a part of the biological food system. The system suggests that as soon as the deficits are resolved, the eating should stop because a stage of satiety is reached. Ideally, this system should regulate food intake and energy balance. It was designed to do so.

WE LIKE WHAT WE DO NOT NEED

The other center of the brain that is involved in the reconstruction of flavor is the orbitofrontal cortex. Compared to the amygdala, the functioning and orientation of this part of the brain is quite different. It is where pleasure or reward is created. In the last 50 years, a lot has changed in the world of food: it has become abundant and basically safe in the Western world. We are generally in a state of good health, abundance, and prosperity, and our interest has moved to the pleasure side of food. We eat for fun and share with friends. The level of pleasure is associated with the release of dopamine. In neurobiology, this neurotransmitter is widely recognized as being critical in our behavior, including in addiction.

With this shift from areas in the brain, the palatability of our food and drinks has gained importance. After all, if there is a choice, why not choose the foods and drinks that we enjoy the most? Consequently, the palatability becomes more important and with it aesthetic values, cognition, and experience. In other words, we are more susceptible to information and thus manipulation by all kinds of advertisements and the opinion of experts and critics. In the best cases, they help you discover and experience elements of flavor that you did not know before.

How about appetite and satiety if eating is done for pleasure? Neuroregulators other than NPY stimulate food intake. They often belong to a family of endogenous opioid peptides that are related to the so-called "reward value" of palatable food, which implies that these peptides are involved in hedonic induced feeding, registered in the orbitofrontal cortex. This system has its own satiety, which is different than biological satiety and is more related to the specific food that is eaten (or beverage that is drunk). Therefore, it is called sensory specific satiety and defined as "the pleasantness of the sight and the flavor of a food eaten to satiety decreases while other foods not eaten to satiety remain relatively pleasant."[23] This basically implies that flavor richness can indeed lead to satiety. Perhaps the best example of that can be seen in the world of cheese.

The difference between the two satieties is clear: satiety itself is based on internal signals, while sensory specific satiety is—at least partly—based on external signals. It has been shown that sensory specific satiety is strongly dependent on sensory aspects of food, such as its flavor and mouthfeel. Take a simple young cheddar or Gouda and a blue vein cheese such as Roquefort. It is safe to assume that most people are likely to eat much more of the simple cheese. With flavor rich cheeses, a relatively small quantity leads to sensory specific satiety. In our modern society with inexpensive, good tasting but not very complex foods that are readily available, many people eat too much.

Because satiety takes time to develop, it can be assumed that slow eating may tend to reduce food intake. On the other hand, quick eating and variety in a meal will enhance intake.[24] In our sudden shift from biological need to eating for pleasure, evolution is too slow to help us. We have to control our intake on our own and, considering the food-related health problems of modern societies, we are not very successful yet in doing so. In government and industry, people are aware that something needs to be done. The big question is how to solve the problems we have created.

LIKING AND THE GUT

The more research in this field progresses, the more secrets are revealed on the receptor level and with regard to neural processing. It becomes clear that we need to include a third party in the understanding of flavor perception and liking: the gut. Everywhere in the stomach, intestine, and pancreas receptors have been identified that detect bitter, sweet, and umami. In the list of cranial nerves that are involved in flavor perception, the vagus nerve is the one to watch. It relates the gut to the brain. The chemosensors in the gut have been linked to appetite and satiety, and to mood and behavior.[25]

The enteric nervous system, as it is called, principally controls the gastrointestinal system. Its role is so important that it is referred to as "the second brain," with a nervous system comprising some 500 million neurons. It has special proteins that are involved in communication, thinking, remembering, learning, and in our mental and physical well-being. It operates independently from and together with the brain. It also uses the same neurotransmitters as the brain. The brain and the gut are involved in appetite and food choice. In fact, about 90% of the signals that are transmitted through the vagus nerve do not come from above but from down under. Indeed, the stomach does a part of the thinking and, in food choice, "gut feeling" may well be a voice to listen to.[26]

Two neurotransmitters that the brain and the gut share are dopamine and serotonin. Dopamine is involved in pleasure and reward; serotonin makes people feel good, prevents depression, and regulates (among other things) appetite. It is assumed that the neurotransmitters that are produced in the gut cannot get into the brain, yet they can affect brain functions and influence a person's mood. Stress is an example. When a person is stressed, the gut increases the production of a hormone called ghrelin. It reduces anxiety and feelings of depression and also stimulates feeling hungry. It does so by stimulating the release of dopamine in the brain. Ghrelin is considered to be the counterpart of leptin, the hormone that is meant to make us stop eating. The combination of stress and the easy availability of energy-dense food may well be related to obesity. The French saying *l'appétit vient en mangeant* (appetite comes while eating) appears to be very true.

Besides its known functions in the metabolism of food, the "second brain" also is involved in developing food preferences, or appetites. The gut-brain pathways are not fully understood yet, and the role of the vagus nerve should deserve more attention to enhance our understanding of human food preference learning.[27]

From this, the conclusion may be drawn that specific neurological processes influence the human motivation to eat and the consequent food choice, and that these processes can have a biological or a hedonic origin. Does this imply that humans automatically like what they want? Berridge reports that it seems axiomatic that we want the rewards we like and like the rewards we want. Yet, although this is usually the case, wanting and liking do not always go together. Research has shown that dopamine-depleted rats still "like" rewards and also still know the rewards they "like," but they fail to "want" the rewards they "like."[28,29] In relation to gastronomy, liking corresponds more closely to palatability, yet wanting is associated with appetite. It is the distinction between the disposition to eat and the actual pleasure of eating, or—in other words—the difference between motivation and affect.

DEVELOPMENT OF PREFERENCES

The basic neural architecture of each sensory system happens in the last 15 to 18 weeks of the development of the fetus and up to the first five months of the newborn's life.[30] After that, the brain keeps on developing, and people become eternal students. Our brain is a big learning machine that constantly stores information. We get older and wiser, and food preferences build up in people. In his book, *Neurogastronomy*, Gordon Shepherd[31] distinguishes the "six ages of flavor."

- It all starts in the womb. A mother's food preferences are passed on. Many studies report examples of how prenatal food experiences lead to greater acceptance and preferences in the newborn. This implies that cultural preferences are more than just eating patterns: you are born with them.
- The influence of the mother continues in the infant stage. The brain is far from fully complete and new cells and connections

keep developing. Breast milk is a carrier of flavor. And gradually the quest for new foods and drinks starts. The brain training begins, including the development of early preferences. Clearly the infant is still dependant on the parents, and thus it helps when they confront their precious toddler with a good variety of flavors.

- The third stage is childhood. Flavor has a powerful effect on young children. It is supposed that they live in a different sensory world than adults. They seem to prefer intense sweet, sour, and salty tastes. Brain systems are still vulnerable, and overeating such foods may lead to disturbance of brain control systems. Soft drinks and fast food are designed to be liked by children in order that they become customers for life.

- Then comes the stage of adolescence. Young people evolve from being largely dependent on their parents to a growing level of independence and start making their own decisions in food choices and in spending money. In the same period, the cognitive level of the brain is still developing. Successful advertising campaigns are targeted specifically to create desire and new preferences.

- In the adult world, new dimensions in food choice come about. Cognition and status become more important. People have to cope with abundance and to control food intake. Many people are conscious about their food behavior and food-related health issues and struggle to control their intake.

- In our modern world, people grow older than ever before. There is an inescapable decline of sensory capacity that is different from person to person. Furthermore, it is hard to assess if, and to what degree, flavor perception has diminished. As long as people care for themselves and prepare their own food, they are likely not to notice the changes at all; they just automatically spice up the food to their liking. This dramatically changes when people become dependent on other people for their care. Not liking the food in nursing homes or hospitals could be solved by paying more attention to flavor and by giving more choices. From a gastronomy point of view, food is considered an investment in the quality of life, and not as a cost.

EXPECTANCY AND LIKING

Flavor, tasting, and language have an intimate relationship. After all, we need words to express ourselves. The flavor experience in the brain needs to be labeled with words. These words in turn are often loaded with personal beliefs and previous experiences and, therefore, they create a certain expectancy. What happens when words are used that are in marked contrast to the actual flavor characteristics? In an experiment, smoked salmon ice cream was used to test this. First, it was named what it factually was: ice cream. In the test, it was strongly disliked. Apparently, people expect ice cream to be sweet and fruity. When it was labeled as a frozen savory mousse, it was accepted. This shows the power of words and the importance of using the right words in the right context. There was another result as well: under the ice cream condition, it was also rated as being more salty than when labeled as a savory food.[32]

Flavors are also related to expectancy in another respect. The odor of a cheese can, for instance, be accepted in relation to the cheese but not when it is taken out of context. This was tested with isovaleric acid. In several trials, it was alternatively labeled as cheddar cheese and body odor. When the odor was presented as body odor in combination with cheddar cheese, it was rated as significantly more unpleasant than when labeled cheddar cheese. The implication is that cognitive factors can have profound effects on our liking foods.[33]

These findings have important implications for gastronomy professionals. Preparing a perfect meal or serving the best wine is not enough. Flavor needs to be explained and presented to the customer in comprehensible words and without raising false expectations. This supports the hypothesis that people in service need to be very well trained and in close contact with the products that they serve. Furthermore, it is wise to pay attention to how dishes are formulated on the menu.

EXTERNAL INFLUENCES ON LIKING

Humans are a part of their environment. What about the influence of season, climate, and weather—can they have an impact as well? The results of an experiment in the Netherlands held in October 2000 suggest that indeed they can (Klosse, unpublished data). Five independent groups of subjects (nonprofessional volunteers, total $N = 238$) across the country were first asked to rate the flavor of a mousse of smoked salmon and a tartare of fresh salmon (summary of the results: Table 7.1).

The mousse was clearly more coating in mouthfeel and rich in flavor, with ripe flavor tones. The tartare was less coating, and the flavor richness was somewhat lower and fresher. In terms of flavor styles, the mousse could be classified as "full" (Segment 6 of the flavor styles cube), while the tartare could be characterized as "fresh" (Segment 4 of the flavor styles cube). Both dishes scored relatively high on palatability (7.8 and 7.3, respectively, on a 10-point scale). When asked which of the two they would prefer on a hot day on a terrace, 89% of the subjects chose the tartare. Keep in mind that this question was asked after the two versions of salmon were tasted and in October. Apparently, people have a strong feeling about when and in what situations certain dishes are appreciated. Consider as well that the mousse was rated higher in palatability.

TABLE 7.1
Sensory Judgment of Two Salmon Dishes

Sensory Judgment ($N = 238$)		Mousse Smoked Salmon	Tartare Fresh Salmon
Flavor Profile:	Mouthfeel contracting	– –	+
	Mouthfeel coating	++	–
	Flavor richness	+	0
	Fresh flavor tones	–	+
	Ripe flavor tones	+	–
Palatability (scale 1–10)		7.8	7.3
My choice for a hot day on a terrace (N)		26 (11%)	212 (89%)

– –: not at all present; – hardly present; 0 indifferent; + present; ++ very present

If this experiment would have been about milk and ice cream this result would not surprise anyone. The "fresh" ice cream would be a logical choice for a hot day on a terrace; a glass of "full" milk is not appealing. A tartare of salmon can hardly be considered to be such a logical and well-known choice. Nevertheless, people are very clear in their preferences, which leads to the conclusion that objective product properties are at the origin of these preferences. Apparently, the tartare of salmon and the ice cream have things in common, making them instrumentally preferable for a hot day on a terrace. The common denominators of ice cream and the tartare are their contracting mouthfeel and freshness. Our conclusion from this experiment is that people know instinctively that the fresh flavor style would suit them best on a hot day. Apparently, liking is not just a function of product characteristics. Flavors need to fit the situation as well. Flavor styles have a function in this respect because they communicate about the flavor to be expected.

PALATABILITY AND THE AESTHETICS OF GASTRONOMY

We have not yet discussed the side of taste that many people often think of—having taste—such as in the ability to distinguish good quality, which is all about recognizing what is aesthetically excellent or appropriate. Aesthetics is the branch of philosophy that deals with the nature of beauty, art, and taste. More specifically, it refers to the creation and appreciation of beauty. Things of beauty are critically reflected upon in an effort to answer the basic question: What makes them beautiful? The answer to this question may then be used to formulate principles for design and to help distinguish art from kitsch, real from fake. Thinking and philosophizing about beauty and aesthetics adds an extra dimension to sensory registration; it gives meaning and value to the simple perception of sensory signals.[34] Therefore, aesthetics has been studied in many sensory fields except where all signals come together: in gastronomy.

Aesthetics is mostly considered to be related to art; it is has been widely studied and for ages. Now, I am neither suggesting that food or beverages are works of art, nor that eating is an artistic activity. Eating and drinking can even be rather vulgar, especially as far as the biological function is concerned. Food from the best chefs in luxury restaurants comes closest to art, just as haute couture does in clothing. Aesthetic principles can be applied in the creation of a dish, the combination of flavors, the colors, and the choice of tableware. Furthermore, beauty is not just an ordinary judgment: it evokes pleasure and desire. In the extreme, people are even willing to pay fantastic amounts of money for a meal or a bottle of wine, far more than the cost to produce it. The same applies to a beautiful piece of art, for that matter.

The reason to pay attention to the aesthetics of gastronomy is not to sell foods and wines as expensively as possible. Knowing more about the beauty of flavor is an intriguing and useful field of study because it may ultimately contribute to the satisfaction of the consumer by our serving foods and wines that are well designed and go well together.

The influential Dominican priest St. Thomas Aquinas (1225–1274), *Doctor Universalis*, formulated in his *Summa Theologiae* that beauty needs three qualities: integrity, harmony, and radiance.[35]

Integrity is the opposite of hypocrisy and has everything to do with honesty, principles, and consistency. Harmonious colors and sounds are pleasing to the eye and to the ear. The opposite of harmony is chaos; our brain has trouble organizing the data and therefore rejects it. In the section "Harmony and Contrast" in Chapter 6, it was stipulated that harmony is an important aspect of food pairing. It helps the brain to quickly process information. Radiance is like the string of a violin; it needs to be set in motion, otherwise nothing happens. Music or things of beauty have the capacity to actually touch you and to generate pleasure. Radiance is therefore considered to be the most important of the three.

This view may well be close to the mark. How can integrity, harmony, and radiance be applied to modern gastronomy? In flavor, tasty is what beauty is in vision and sound. It is the capacity of foods and drinks to please our brain. We have seen before that, to achieve this, our brain needs to be excited. To do that, we have to organize flavor signals in such a way that our brain can quickly process them and place them in a harmonious order. The study of how the brain responds to "beautiful" sensory signals is called neuroesthetics. It is suggested that neurology will eventually uncover laws of aesthetic experience that identify the common preferences for symmetry, grouping, and proportion that successful artists apply intuitively.[36] For the time being, gastronomy is not specifically included in neuroesthetics. Research is needed to see if the same principles apply. For now we can safely conclude that, in matching food and beverages, harmony can also lead to liking, provided that the combination is not boring.

In her book *Making Sense of Taste*, Carolyn Korsmeyer[37] elucidates beautifully that although food—or rather the creation of flavor—is not an art in itself, there are dimensions that make fine food and wine comparable to works of art. Beauty is not an accidental experience that just happens to someone haphazardly; it is primarily the result of doing things right and requires cognition. Clearly, as we have seen, pleasure is not guaranteed. Appreciation for works of art first requires a certain understanding and insight. More is needed: our brain needs to become aroused. Knowledge, experience, expectation, and elements of surprise are required for this to happen. Although these elements are directly related to the taster, they can be influenced. In hotels and restaurants, they can be managed; or should I say *must* be managed? A well-trained staff in the kitchen and in service can do wonders when they are aware of their role in arousing pleasure. Besides making the right choices in composition, they can provide the guest with functional information, providing the necessary insights that are essential in generating expectation and in organizing surprise.

FLAVOR STYLES REVISITED

The flavor styles are now introduced again and in a practical setting. They can be functional in daily practice, but this also raises the question: What do consumers have as a basis for making food choices? The flavor of foods or drinks is generally described based on:

- Product category (e.g., meat, fish, beer, coffee)
- Ingredients (e.g., filet of beef, salmon, grape variety)
- Origin (e.g., Brie cheese, Bordeaux wine)
- Brand name (e.g., Coca-Cola, Heineken, Mars, Heinz Tomato Ketchup)

Except for the last category, the problem with these generally used notions is that they say very little about the flavor of a specific product within a certain category. In other words: wines coming from the Bordeaux region do not have the same flavor, and neither do all wines made from the Chardonnay grape taste similar nor do all salmon dishes or Brie cheeses, to mention just a few examples. With regard to flavor, brand names only serve a purpose for people who are acquainted with the brand from previous experiences.

Broadly defined product categories without further specification run the risk of being categorically disliked if one or a few products within that category are not liked. In a study conducted by Wardle et al.,[38] such broad categories emerged in a factor analysis of food preferences in young children. Children tended consistently to like or dislike foods within each of these factor categories. Such preferences could not be linked to simple sensory preferences such as sweetness, saltiness, or creaminess. For instance, many sweet foods (e.g., chocolate) were not associated with the factor "dessert," and many savory foods (e.g., sausages and burgers) did not load in the meat and fish factor. The authors suggested that these results support the notion that foods are recognized as a synthesis of many sensations, including taste, olfaction, and aspects that are involved in mouthfeel (texture, fluidity, and temperature). Foods can have very different profiles and, therefore, broad categories are hardly suitable as a basis for identification as is illustrated in this study. Because flavor styles are based on this so-called "synthesis of many sensations," a different structure of preferences is likely to emerge—one that is not broadly defined but more specific. Perhaps some extrinsic factors can play a role. A gift as a reward for food choice has already been successfully applied by one multinational restaurant chain promoting "Happy Meals."

At present, most food labels do not give the information that is required to enable consumers to compare and select foods on the basis of flavor properties.[39] Flavor style labeling could solve at least part of this problem. Take wines, for example. Wine labels traditionally feature the regional origin and/or the grape variety as identification. Both are poor indicators of the flavor, which can therefore be quite different from that which is expected. Potentially, this will lead to disappointment that can be prevented by mentioning the flavor style of the wine on the label. The flavor style would enable the consumer to select the wine that suits his preference, thus giving him the opportunity to make the right choice.

To see how this can be done, go back to the aforementioned experiment with the salmon dishes and relate the outcome to Chardonnay, for example. Flavor profiles of wines made from this grape variety show differences ranging from fresh to full. It can be expected that Chardonnay wines in the flavor style of "fresh" are fit to drink on a warm day on the terrace, just as—or even with—the tartare of salmon. It is therefore essential for potential consumers to have a good notion of the flavor to be expected—before they make their food choice. Sommeliers need to be trained to be aware of these aspects of wine; and ideally, the dishes of restaurants should also be suited to the time of the year or even the day.

THE THEOREM OF FLAVOR STYLES

In general, expectancies play an important role in liking. The package, label, and advertisements of products are mostly designed to raise a consumer's expectation of a product. Care should be taken that the product meets these expectancies. If it does, it is reported to be able to modify people's actual perceptual experiences, similar to placebo effects in medicine. The magnitude of this effect is related to the availability and reliability of sensory information. If there is ample reliable information, the role of expectancy is smaller than when sensory information is less available and less reliable.[40]

Flavor styles can prove to be particularly useful in those cases where the amount of reliable sensory information is low. This applies especially in product categories where the flavor of individual products varies strongly within the category and to products that are unfamiliar to the consumer or products of which the flavor can hardly be assessed prior to tasting (i.e., new products, new wines, and new prepared dishes). Flavor styles provide reliable sensory information for these kinds of products. Well-known brands have little need for a flavor style as identification. The flavor is always the same, and the brand name itself serves as identification. Based on these observations, the following theorem of flavor styles can be formulated: the value of flavor styles

FIGURE 7.3 THE THEOREM OF FLAVOR STYLES.

increases as the amount of reliable sensory information decreases. Figure 7.3 illustrates this.

THE NEW PARADIGM

In view of all this, the question must be raised of whether the paradigms that are generally assumed are still valid. Considering everything that we have disclosed about gastronomy, the conclusion must be that it is a holistic science. The Greek word *holos* means whole, total, all. The concept of holism suggests that the functioning of natural systems cannot be fully understood by studying their component parts. Holism makes some people frown. Therefore, it is good to specify its meaning in the context of gastronomy. For us, it is the integration of the classical, reductionist scientific principles and other, less rigorous approaches. Mainstream, exact, and reductionist science has undoubtedly proven its value, especially in the world of the natural sciences, which was developed in the sixteenth and seventeenth centuries, inspired by the work of René Descartes. In essence, the great geniuses of those times embraced the idea that any phenomenon, including the whole universe, could be fully understood by reducing it to its essential building blocks. In the quest for understanding, matter is broken down into ever smaller bits. The latest triumph in this respect was the discovery of the Higgs particle.

Gastronomy is defined as the science of flavor and tasting. This implies that it is by definition a combination of natural sciences (flavor) and human-related sciences (flavor perception). As we have seen, reductionism rules in the world of flavor, yet it fails to explain what happens in the world of flavor perception. The reductionist focus on the basic flavors has always dominated the science of flavor; in other words, the paradigm used was far from complete. And even if all senses that are involved in the perception of flavor were included, the role of the brain and the gut needs to be included.

Similarly, in the world of food, focus has shifted from whole foods to the nutrients they contain. The word *nutritionism* has been introduced to indicate this shift from real foods to the nutrients they supposedly contain.[41] The food industry has exploited these ideas to the max. The positive or negative health effects of our food intake are advertised more and more. The inescapable result is that the control of the food industry over our food choices has increased. Gastronomy can be considered as the antonym of nutritionism because it is focused on the whole experience, including the liking and enjoying of food and drink. The holistic perspective of gastronomy looks at the food-related problems of modern Western societies from a different angle. It focuses not solely on the food and its composition but on the human that eats it as well. And, in doing so, gastronomy is especially interested in why this human likes the things he eats or drinks. However different from today's approach, this actually implies going back to the roots. The father of Western medicine, Hippocrates, was convinced that illnesses were the result of environmental factors, diet, and living habits. One of his famous quotes is "let food be thy medicine and medicine thy food." And another one, "it is far more important to know what person the disease has, than what disease the person has."

The reductionist, evidence-based approach in medicine today has been very successful. The result is a thorough insight into the cause and often the cure for many medical problems. However, research on the virtues of whole foods with regard to health is difficult. This gives some people reason to rely on proven pills rather than on unproven diets. Nevertheless—and luckily—the advice to eat at least moderate amounts of fruits and vegetables is still maintained.[42] In view of shifting to another paradigm, accepting practice-based evidence could be a step forward but clearly under the condition of using the best methods and scrutinizing results.

IS FOOD THE STIMULUS
OR THE RESPONSE?

In 1904, Ivan Pavlov was awarded the Nobel Prize for showing that not the actual food but the thinking about it, triggered by the sound of a bell, induced the physiological reflex of salivation. Therefore, Pavlov is supposedly the first to have demonstrated that food choices are not only motivated by product values or biological needs. His experiment showed the importance of cognition, which was later referred to as classical conditioning. In psychology, the principles of classical conditioning (stimulus–response) have developed into behaviorism, a concept that is associated with the work of B. F. Skinner.[43] Humans are individuals and at the same time a part of the environment. The study of behaviors involves the interaction between organism and environment. A person is first of all an organism. In becoming a person, the organism acquires a repertoire of behavior. This framework determines his operant behavior; this is always under control of a current setting. Tasting can be seen as operant behavior that takes place within a certain setting. In fact, on the human side, there are many forces that influence food behavior and liking.

There are many everyday examples. Much of what is liked and disliked is a result of conditioning and/or learning. Coffee, alcohol, tobacco, and some vegetables are well-known examples of products that a person eventually learns, step by step, to appreciate. Often there are underlying needs. Smoking a cigarette is often related to stress; many people drink alcohol as a part of social customs, while coffee is supposed to increase the level of activity and is often related to sleeping difficulties. Some of these popular beliefs may not be true or the products even unhealthy, but that is irrelevant because they act as stimuli. It is the starting point of the taste process.

This line of reasoning moves away from the classical sensorial conception where the product is considered to be the stimulus and tasting as the logical response. The brain is not a passive stimulus-driven device. The processing of stimuli is controlled by top-down influences that constantly create predictions about what is to come. This influences perception. The taster is not a mere observer, (s)he is a part of a process that (s)he helps shape. From that perspective, the product should be considered as response. The brain, our mood or feeling, possible deficiencies, or plainly the situation or availability provide the stimulus for the consumer; the selection of the product is in that case the response (see Figure 7.4). Consequently, to develop a better understanding of liking, the decision to eat or drink should to be taken as a starting point. The actual act of tasting is a consequence of a choice that has more or less consciously been made.

The significance of this turnaround is that the consumer's personal circumstances and expectations are included in the food choice, which consequently reduces the complexity of understanding flavor appreciation. The implication is that the consumer needs to be able to recognize the flavor of products in such a way that (s)he is able to make the right choices. If these products are well made, not spoiled, and true to the expectation, it can safely be assumed that these "right choices" will be appreciated; they fit the individual beliefs and circumstances. For a person to be able to make such choices, flavor has to be made tangible, recognizable, and/or comprehensible. The use of flavor styles—the classification of flavor based on product characteristics—may be more useful than superficially thought.

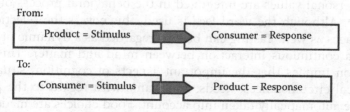

From:

Product = Stimulus ➡ Consumer = Response

To:

Consumer = Stimulus ➡ Product = Response

FIGURE 7.4 THE STIMULUS–RESPONSE TURNAROUND.

WHERE IT ALL COMES TOGETHER: THE BEHAVIORAL MODEL OF FOOD CHOICE

The process of food choice is provoked by either internal or external stimuli. Internal, biological, and physiological forces such as hunger or thirst and more complex signals that stimulate the intake of specific foods have always attracted much attention. However, they cannot explain many aspects of eating behavior.[44] Environmental factors such as visibility, smell, verbal appraisal, or plain availability and convenience can stimulate food behavior.

The human brain plays a central role in the taste process. It is active in all aspects of taste, in the choice of the substance to be eaten or drunk, and in the physiological registration of flavor. Aspects such as the time of the day, personal circumstances and preferences, past experiences, culture, physical needs and wants, presence of substances, and so on ultimately will influence the food choice.[45]

Personal food choices are determined by all kinds of environmental and individual circumstances, which implies that in food selection there is not one, rational good choice. Consumers make their best choice, considering the circumstances. Categorizing foods is an important instrument in these personal food systems, which raises the question of how this is done. Flavor styles can be particularly useful in food identification, which facilitates categorization and may, therefore, prove to make food choices more in line with expectation.

What is tasted is the outcome of some kind of decision-making process that preludes flavor registration. "Liking" implies being satisfied about the choices that have been made. Consequently, the perception of quality must also be closely related to this process. Product and personal values are integrated in the behavioral process of food choice. Although the word *food* is used, this covers the selection of beverages as well. Tasting can best be regarded as a dynamic process with a continuous interaction between mind and matter. This orientation implies that the important aspects of cognition, situation, physical circumstances, needs, and wants with regard to the product are automatically taken into account. Food choices are made and expectation is set. In the actual tasting, the result is evaluated. It is highly likely that the brain will pass a favorable judgment (liking, quality) on those nutrients that fulfill or exceed the set expectation. The human mind files these results and as such they will be a part of the next round of decision making. The perception of product characteristics and product quality is influenced by these factors. This process is illustrated in Figure 7.5.

This process shows that flavor registration takes place within a frame that is shaped by biological, psychological, economic, social, and cultural forces. The process of food choice commences within the brain and is stimulated by either internal (e.g., hunger, thirst) or external stimuli (e.g., smelling, seeing, hearing about potentially attractive nutrients). Then, the following stages can be distinguished in the process:

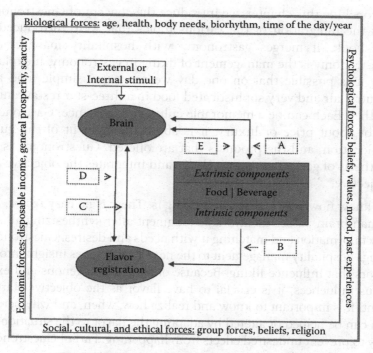

FIGURE 7.5   THE BEHAVIORAL MODEL OF FOOD CHOICE.

A. Yielding to a stimulus, which implies food choice and the selection of a food product. Information goes from mind to food and from food to mind. It is important to note that the food is first selected based upon its extrinsic components (e.g., package, smell, name). These raise expectation.
B. The actual tasting and flavor registration.
C. Evaluation in the human brain: Does the food live up to expectation and does it satisfy needs or wants?
D. The result of this evaluation leads to acceptance or rejection, and may influence the pace of eating.
E. The filing of information about the food, cognition, which becomes a part of the next food choice processes.

INTRODUCING: GUESTRONOMY

To conclude this chapter, we introduce the concept of Guestronomy. It is the integration of the various topics that were introduced in this book. It merges gastronomy with hospitality management. Guestronomy is the management of liking in gastronomy. It explains how it is possible that on one day we can enjoy simple food at a country fair and very sophisticated food in a three-star restaurant on another. Each can be a memorable culinary experience. Gastronomy is not about price or luxury, it is about the right fit of the guest, the location, and the products that are offered. Guestronomy is the synthesis of gastronomic knowledge and integrates the objective and subjective.

We know now how complicated liking is. There is no easy recipe. The human brain passes its favorable judgment after synthesizing all sensory information and matching it with needs and desires. Guestronomy brings hospitality management to the next level. It gives insight into the factors that influence liking. Because of all the endogenous and exogenous influences, it is crucial to have flavor as the objective starting point. It is important to know and realize how, when, and why perception can be helped or altered by adapting the flavor to the situation—if only to prevent undesired effects from happening. Part of Guestronomy is to take these effects into account.

	Material	Personal
Fixed	Ambiance (hardware, situation, decoration, music)	Personality, frame of mind, experience, cognition
Variable	Products: food and beverages	Hospitality and timing

FIGURE 7.6 THE MATRIX OF GUESTRONOMY.

The matrix shown in Figure 7.6 was created based on the parameters of material–personal and fixed–variable. The four separate quadrants that emerge from this mix are separate fields of interest, each with its own dynamics. This shows that there is not one good answer in gastronomy. The best experiences are a result of a perfect match of the different factors and a perfect fit of all quadrants. In the gastronomy approach, flavor is defined as a product quality. This implies that it can be manipulated to fit the other quadrants. A simple dish that fits the situation can be liked more than a refined one that does not fit. In hospitality, this needs to be understood and recognized; this requires flexibility as well as practical knowledge that can immediately be applied, if necessary.

We will explain this by introducing the quadrants.

1. Material:Fixed. This is the world where the facilities are—the situation. A castle, a boat, a fast-food restaurant, in a city, close to the sea, on a farm, or in a skyscraper, every situation calls for a certain expectation and has its limitations and opportunities. Some of these elements, such as lighting or music, can be changed more easily than others. Nevertheless, when a guest enters a restaurant it is a fixed situation that is not easy to adapt to individual preferences.
2. Personal:Fixed. This is the world of the guest, with his likes and dislikes, experience, and culture. He has made his choice of a

certain location and waits for the great experience—the one that will make him happy and give him that lasting memory.

3. Material:Variable. This is the world of the products, foods, and beverages; let us consider them as the materials that can make the guest happy. They have to be up to standard and fit with each other and with the rest of the quadrants. Ideally, the right food is selected or recommended to fit the situation and the person.

4. Personal:Variable. This is the world of hospitality, reading the guest and the situation. Trying to suggest the right products for the situation and to serve them in the desired time frame.

SUMMARY

Flavor can be studied from a human and a product angle. In this chapter, the synthesis was reviewed. What happens after the registration by our sensorial systems? How does the information converge, and what can we say about palatability and liking? Palatability is defined on the product level and, therefore, it belongs to the objective domain. We have shown that earlier by formulating the culinary success factors. In this chapter, the neural aspects were introduced—neuroesthetics, for instance, and the role of language, expectancy, and cognition. The flavor styles are functional in communicating the flavor. Liking is all about a favorable judgment about what is eaten or drunk. It has to do with replenishing deficiencies (we like what we need) and with plain pleasure, which is the result of the release of dopamine.

Differences between humans and cross-modal sensorial interferences influence registration and tasting. If the physiology underlying the transduction of single sensory modalities is already complex, the physiology underlying a multi-sensorial experience such as tasting is at least three times as complex. The four basic tastes hypothesis has dominated the field and research methods. This implies that attention has been focused on the gustatory part of flavor registration. Consequently, answers on flavor registration as a whole are not to be expected.

Gastronomy is a holistic science in the sense that it endeavors to integrate the virtues of reductionism without losing the vision of the whole. Guestronomy is introduced as integrating flavor and flavor perception, the two sides of the same coin: what it is and how it is perceived.

NOTES

1. Beidler, L. M. (1978). *Biophysics and chemistry of taste*. In Carterette, E. C. and Friedman, M. P. *Handbook of perception*, Volume VIA, Tasting and Smelling (pp. 21–49). New York: Academic Press.
2. Spector, A. C. (2000). *Linking gustatory neurobiology to behaviour in vertebrates*. Neuroscience and Behavioural Reviews 24:391–416.
3. Smith, D. V., Margolskee, R. F. (2001). *Making sense of taste*. Scientific American 288:26–33.
4. Smith, D. V., St. John, S. (1999). *Neural coding of gustatory information*. Current Opinion in Neurobiology 9:427–35.
5. Gilbertson, T. A., Damak, S., Margolskee, R. F. (2000). *The molecular physiology of taste transduction*. Current Opinion in Neurobiology 10:519–27.
6. Rolls, E. T. (2005). *Taste, olfactory, and food texture processing in the brain, and the control of food intake*. Physiology & Behavior 85(1):45–56.
7. Bartoshuk, L. M. (1978). *History of research on taste*. In Carterette, E. C. and Friedman, M. P. *Handbook of perception*, Volume VIA, Tasting and Smelling (pp. 3–16). New York: Academic Press.
8. Van den Akker, E. H. *Een kwestie van goede smaak, chorda tympani en middenoorchirurgie*. Referaat 25 mei 2000, Kliniek Keel-, Neus- en Oorheelkunde, Universitair Medisch Centrum, Utrecht. (in Dutch)
9. Lehman, C. D., Bartoshuk, L. M., Catalanotto, F. C., Kveton, J. F., Lowlicht, R. A. (1995). *Effect of anaesthesia of the chorda tympani nerve on taste perception in humans*. Physiology & Behavior 57:943–51.
10. Winquist, F., Wide, P., Eklov, T., Hjort, C., Lundstrom, I. (1999). *Crispbread quality evaluation based on fusion of information from the sensor analogies to the human olfactory, auditory and tactile senses*. Journal of Food Process Engineering 22:337–58.

11. Clydesdale, F. M. (1993). *Color as a factor in food choice*. Critical Reviews in Food Science and Nutrition 33:83–101.

12. Morrot, G., Brochet, F., Dubourdieu, D. (2001). *The color of odors*. Brain and Language 79:309–20.

13. De Lange, C. (2012). *Feast for the senses*. NewScientist 216(2896/2897):60–2.

14. Shimojo, S., Shams, L. (2001). *Sensory modalities are not separate modalities: plasticity and interactions*. Current Opinion in Neurobiology 11:505–9.

15. Calvert, G. A., Campbell, R., Brammer, M. J. (2000). *Evidence from functional magnetic resonance imaging of crossmodal binding in the human heteromodal cortex*. Current Biology 10:649–57.

16. Oberfeld, D., Hecht, H., Allendorf, U., Wickelmaier, F. (2009). *Ambient lighting modifies the flavor of wine*. Journal of Sensory Studies 24:797–832.

17. Painter, J. E., Wansink, B., Hieggelke, J. B. (2002). *How visibility and convenience influence candy consumption*. Appetite 38:237–8.

18. Spence, C., Shankar, M. U. (2010). *The influence of auditory cues on the perception of, and responses to, food and drink*. Journal of Sensory Studies 25:406–30.

19. Oka, Y., Butnaru, M., von Buchholtz, L., Ryba, N. J. P., Zuker, C. S. (2013). *High salt recruits aversive taste pathways*. Nature doi:10.1038/nature11905.

20. Poothullil, J. M. (1995). *Regulation of nutrient intake in humans: a theory based on taste and smell*. Neuroscience and Biobehavioral Reviews 19:407–12.

21. Bray, G. A. (2000). *Reciprocal relation of food intake and sympathetic activity: experimental observations and clinical implications*. Int. Journal of Obesity 24(suppl. 2):S8–17.

22. Shiraishi, T., Oomura, Y., Sasaki, K., Wayner, M. J. (2000). *Effects of leptin and orexin-A on food intake and feeding related hypothalamic neurons*. Physiology & Behavior 71:251–61.

23. Rolls, E. T., Rolls, J. H. (1997). *Olfactory sensory-specific satiety in humans*. Physiology & Behavior 61:461–73.

24. Bray, G. A. (2000). *Afferent signals regulating food intake*. Proceedings of the Nutrition Society 59:373–84.

25. Trivedi, B. P. (2012). *Neuroscience: hardwired for taste*. Nature 486:S7–9.

26. Young, E. (2012). *Alimentary thinking*. NewScientist 216(2895):38–42.

27. Sclafani, A., Ackroff, K. (2012). *Role of gut nutrient sensing in stimulating appetite and conditioning food preferences*. Am. J. Physiol. Regul. Integr. Comp. Physiol. 302(10):R1119–33.

28. Berridge, K. C. (1996). *Food reward: brain substrates of wanting and liking.* Neuroscience and Biobehavioral Reviews 20:1–25.
29. Berridge, K. C., Robinson, T. E. (1998). *What is the role of dopamine in reward: hedonic impact, reward learning, or incentive salience?* Brain Research Reviews 28:309–69.
30. Graven, S. N., Browne, J. V. (2008). *Sensory development in the fetus, neonate, and infant: introdution and overview.* Newborn and Infant Nursing Reviews 8(4):169–72.
31. Shepherd, G. (2012). *Neurogastronomy. How the brain creates flavor and why it matters.* New York: Columbia University Press.
32. Yeomans, M., et al. (2008). *The role of expectancy in sensory and hedonic evaluation: the case of smoked salmon ice-cream.* Food Quality and Preference 19(6):565–73.
33. Rolls, E. T. (2005). *Taste, olfactory, and food texture processing in the brain, and the control of food intake.* Physiology & Behavior 85(1):45–56.
34. Zangwill, N. (2007). Aesthetic judgment. *Stanford Encyclopedia of Philosophy, 02-28-2003/10-22-2007.*
35. Original text: *ad pulchritudinem tria requiruntur, integritas, consonantia, claritas.*
36. Reber, R., Schwarz, N., Winkielman, P. (2004). *Processing fluency and aesthetic pleasure: is beauty in the perceiver's processing experience?* Personality and Social Psychology Review 8(4):364–82.
37. Korsmeyer, C. (1999). *Making sense of taste; food and philosophy.* Ithaca, NY: Cornell University Press.
38. Wardle, J., Sanderson, S., Gibson, E. L., Rapoport, L. (2001). *Factor-analytic structure of food preferences in four-year-old children in the UK.* Appetite 37:217–23.
39. Monro, J. A. (2000). *Evidence-based food-choice: the need for new measures of food effects.* Trends in Food Science and Technology 11:136–44.
40. Dougherty, M. R. P., Shanteau, J. (1999). *Averaging expectancies and perceptual experiences in the assessment of quality.* Acta Psychologica 101:49–67.
41. Pollan, M. (2008). *In defense of food. An eater's manifesto.* London: Penguin Books.
42. Key, T. J. (2010). *Fruit and vegetables and cancer risk.* British Journal of Cancer 104:6–11.
43. Skinner, B. F. (1974). *About behaviorism.* New York: Knopf.
44. Levitsky, D. A. (2002). *Putting behavior back into feeding behavior: a tribute to George Collier.* Appetite 38:143–8.
45. Drewnowski, A. (1997). *Taste preferences and food intake.* Ann. Rev. of Nutrition 17:237–53.

EPILOGUE

"Prediction is a risky business, especially if it is about the future." What about the future of gastronomy? The concept of flavor styles constitutes a new paradigm and places gastronomy in the realm of science. The mission of gastronomy is to attain a better understanding of liking, and gastronomy's future depends on the extent to which the new paradigm succeeds in doing that. It all starts with the assumption that liking is not a coincidence. This makes it interesting and useful, and in some cases even exciting, to know more about it.

Cognition is defined by the *Oxford Dictionary* as "the mental action or process of acquiring knowledge and understanding through thought, experience, and the senses." The word *cognition* comes from the Latin word *cognoscere* ("getting to know"). In it is the Greek word *noesis*. It means understanding and is a conjunction of *noein* ("to perceive") and *nous* ("mind"). Considering this, have you ever wondered about the word *recognition*? If you recognize something, it must be something that you have previously learned—a concept that you have become familiar with. Otherwise you are not able to recognize it. The human brain needs concepts to label perceptions. Knowledge paves the way for recognition. Anything can only be understood with the knowledge that you have.

Gastronomy is no exception. It is a constant process of learning. Every bite or sip you consciously take enhances your knowledge. It is learning by doing, and in this specific case, we do not even have a choice because we need to eat and drink. Even so, our brain needs concepts to recognize what we taste. I am inclined to believe that the lack of systematic study and training made flavor and everything that relates to it hard to recognize. By introducing new concepts and rearranging existing knowledge, new insights have emerged. These new concepts enable the recognition of other facets of flavor and tasting in a way that people can comprehend and learn rather easily.

Organisms use senses to gather information about the outside world. The senses are often astoundingly sophisticated and help organisms to survive, for instance, by guiding them to food and warning them about potential dangers. In humans, these systems are not as well developed as in most animals, but they suit our needs, especially because our brain and senses cooperate well together. The human brain has learned to make a lot out of little information. Particularly in tasting, this process is very intricate indeed, and information is swiftly interpreted, as we have seen. Colloquially, many people do not speak about taste but about liking. Therefore, they do not speak about what it is but about whether they appreciate it or not. One may never have learned to speak about flavor and, therefore, have never been trained in its recognition.

The development of science is a fascinating showcase of how our view of the world has gradually evolved by gaining new insights. There may well be parts of the world that we have not yet learned to recognize because we lack the concepts. In his essay "Physics and Reality," Albert Einstein writes, "But even the concept of the 'real external world' of everyday thinking rests exclusively on sense impressions."[1] What is the real world if perception and knowledge vary from person to person? What is reality? One definition of reality says it is "the world without us."[2] It is the world that is untouched by human recognition, desires, and intentions. It is the fundamental distinction between flavor and tasting. The first is real and exists without us, the second leads to our perception of reality. We need good concepts to try to understand reality.

New worlds beyond our senses not only become apparent by defining new concepts. In the physical world, concepts become real as soon as they can be perceived and measured. Before Isaac Newton, people believed that white light was colorless and that the prism produced the colors. This paradigm shifted when Newton's experiments demonstrated that white light actually consisted of the colors of the rainbow, which the prism made perceptible. Subsequently, the visible spectrum was made measurable, and now we know that the wavelengths between

622 nm and 780 nm correspond to the color red. This is purely functional and has no relation whatsoever to our being able to see red, our opinion, love, or to cultural or practical uses that involve red. Yet it truly helps to have scales and measures to classify colors.

In the case of flavor, the question is whether scales and measures will ever be found. For certain, it is extremely hard, and it may be even impossible, to regard flavor purely in the physical world without us, utterly separated from senses and human cognition. We can indeed use biochemical and physical analysis to create as much distance as possible between flavor and human perception. Yet there is another question: are there other methods or approaches that can lead to useful results? The flavor styles model is functional for making the flavor world tangible and intelligible, at least to a certain extent. Its concepts—contracting, coating, and flavor richness—have proven to be valuable descriptors in practice. What is more, they allow flavor to be assessed in natural situations. This is particularly important now that we know that flavor perception must be seen as a sublimation of the input of all our senses. Can fundamental answers be expected from reductionist, deductive reasoning if the whole is more than the individual parts? Surely a dinner at a table, be it in our home or in a restaurant, is very different from having a bite or a sip when imprisoned in a scanner with a head full of electrodes or from tasting food in isolated white booths with infrared lighting.

New insights do not always come easily. It is remarkable that once they are seen everything falls into place, and it is hard to conceive that you did not see it before. That is the virtue of a good theory: it provides a framework that explains the phenomena that we perceive. You need a theory to explain the facts, and the theory stands until facts turn up that call for a critical evaluation and possibly an adaptation or even a novel theory. Nevertheless, however logical and useful theories may be, it may be a while before new theories are accepted. This is not strange. New insights are often the result of libertinism and a break with traditionalist views. They question what is generally believed and, by accepting them, they implicate changing beliefs, teachings, curricula, and so on. Indeed it is much easier to leave everything as it was and to try to squeeze the new facts into the old frameworks. However, that is not the way to progress.

Breaking with traditions is fine if those traditions are replaced by better ones. But how do you know? Can this new gastronomy theory be accepted? Evidently, it would greatly help if there were already objective scales and measures to establish flavor profiles. I suppose that in time scientists will succeed at this endeavor. For the time being, this new theory can be considered the best available map for embarking upon an academic journey. Most likely it is not perfect, but it does give you a good idea where you are heading and what to expect. As with any other expedition, you use your map as a guide, but you keep your eyes open. You never know what you will come across. The map is perfected while traveling. And when you return, your map is much better than when you left. Max Planck might also be right: "a new scientific truth does not triumph by convincing its opponents and making them see the light, but rather because its opponents eventually die, and a new generation grows up that is familiar with it."[3]

NOTES

1. Einstein, A. (1936). *Physics and reality.* Journal of the Franklin Institute 221:349.
2. Westerhoff, J. (2012, October). *What is reality.* New Scientist, Issue No. 2884.
3. Kuhn, T. (1970). *The structure of scientific revolutions.* Chicago: Chicago University Press, p. 150.

INDEX

A

Acetaldehyde, 214
Acidity, 42, 87–88
  of beers, 198, 266
  of fruits, 154–155
  matching wines and food and, 238,
    239–241
Acids, 87–88
  lactic, 216
  malic, 216
Acquired tastes, 257
Adaptation, 43–44
Adenylic acid (AMP), 91, 215
Adolescence, 286
Adrià, Ferran, 114–115
Aesthetics of gastronomy, 289–291
African Americans, 29
Aftertaste, 167
Aging
  of cheeses, 83
  of meat, 124–125
  of wines, 204, 216, 218, 221–223
Agriculture, 115, 116
  organic, 122
Al cartoccio, 137
Alcohol. *See also* Beer; Wines
  cross-adaptation and, 46
Alginates, 106, 142
Altitude, 143
Alzheimer's disease, 37
Amateur tasters, 51–52
Amino acids, 101–103
Amygdala, 275, 280

Anthocyanins, 210, 211, 219
Appellations of Origin, 117
Appetite, 282–285
Apples, 62–63
Aqueous solutions, 177–178
Aquifers, 179
Aristotle, 14, 38, 76
Aromas. *See also* Smell
  in Cabernet Sauvignon grapes,
    207–209
  of fruits, 155
Aromatized tea, 187
Ascorbic acid, 88
Asian cuisines, matching wines and
    food and, 258
Astringency, 41–42, 90
  as form of dryness, 66
  matching wines and food and, 248
Astringent/drying compounds, 40

B

Baking, 133
Balance
  of flavor components, 166
  of flavor styles, 78
Bao technique (frying/pan frying/stir
    frying), 131–132
Barbecuing, 134
Barham, Peter, 9
Basic tastes (primary tastes), 13–14
  problems and misconceptions,
    30–32
  receptors for, 8–9

Bâtonnage, 216
Bean family, 151–152
Beer, 192–198
  alcoholic content of, 196–197
  brewing, 193–195
  differences between wine and,
      226–227
  matching foods and, 265–267
  top-fermenting and bottom-
      fermenting, 195, 196
  types and flavor profiles, 196–198
Behavioral model of food choice,
    298–300
Behaviorism, 296
Bell peppers, influence of preparation,
    173
Berridge, K. C., 285
Beverages, 177–227. *See also specific*
      beverages
  aqueous solutions, 177–178
Bitter flavors, 44
Bitterness, 29
  of beers, 197, 266
  characteristics of, 88–90
Bitter substances, astringency, 66
Bitter taste, 31, 32
Black tea, 186
Blanching, 128–129
Blending wines, 220–221
Blumenthal, Heston, 142
Bocuse, Paul, 138
Brain, 297–300
  tasting and, 272–277
Braising, 135
Breeds, 120, 121
Brettanomyces, 43
Brillat-Savarin, Jean-Anthelme, 5, 8, 44,
    276
  on preparing food, 126
  on ripeness, 123
Brining, 139
Burning compounds, 40
Butter, frying in, 131

C

Cabbage family *(Cruciferae, Brassicaceae),*
    151
Cabernet Sauvignon, 209, 221
Caffeine, 44
Capsaicin, 44
  matching wines and food and,
      249–250
  nociceptors and, 41
Carbohydrates. *See also* Starch
  low-fat diet and, 98
  properties of, 98–100
Carbonation, 14
  of mineral waters, 180, 182
  of wines, 181–182. See also
      Sparkling wines
Carbon dioxide (CO_2), 32, 37, 45
  matching wines and food and,
      246–247
  in mineral water, 179, 180
Carbonic acid, 31, 64, 88, 140, 181,
    266
Carrageenans, 106
Carrot family, 152
Carry-over effects, 46–48
Cassava, 113
Catching Fire (Wrangham), 112
Cattle, breeds of, 120
Cayenne peppers, 42
Celeriac, 153
Celery, 153
Cellulose, 100
  microcrystalline, 106
Cellulose gum, 106
Cephalization, 112
Chao technique (sautéing/stir frying),
    132–133
Cheese
  aging, 83
  matching wines and, 250–254
  odor of, 287
Chemical & Engineering News, 106

Chemical senses, 12–14, 25
  classification of, 13
Chemical stimuli, 24
Chewing, 25
Chianti wines, 200–201
Chicken breast with a creamy sauce and
    tarragon, manipulating flavor
    profile of, 70
Children, 28, 29, 286
  factor analysis of food preferences
    in, 292
Chlorhexidine, 45
Cholesterol, 97
Chorda tympani nerve, 27, 276, 277
Circumvallate papillae, 25–27
Citric acid, 45, 87, 88, 154, 241
Classical conditioning, 296
Clay pot cooking, 137–138
Climacteric fruits, 124
Climate, 118–119
  grape growing and, 202–204
Coagulation, 102–103, 126
Coating, 15, 65–67
Coffee, 188–192
  hard water and, 178
Cognition, 307
Cold cooking, 138–140
Collagen, 102, 103, 105, 107, 121, 135,
    147
Colloid science, 38
Colors, 51, 163
Complexity, 70–72
Conduction, 127
Contracting, 15, 64–67
Convection, 127
Cooking, 128–129
Cooking techniques. See Preparation
    techniques
Cool compounds, 40
Cramwinckel, Ir., 63
Cranial nerves, 276, 284
Crispness, 50
Cross-adaptation, 46

Cross-modal interaction of the senses,
    50, 51, 278–279
Croûte, 137–138
Cryosearing, 141–142
Crystallization, 94
Cucumber family, 152
Culinary success factors (CSFs), 165–170
Cultural context, matching wines and
    food and, 257
Curing, 139
Curry, 149–150
Cuvées, 221

D

Decanting wines, 226
Deep-frying, 136
Dehydration, 140
Demi-glace, 147, 246
De re coquinaria (Apicius), 91
Desensitization, 43–44
Diffusion, 129
Disaccharides, 84, 98
Distilled water, 178
Dominant factor, matching wines and
    food and, 235–236
Dopamine, 284, 285
Dough, for sealing meat, 138
Drugs, flavor registration and, 37
Drying, 140
Dryness
  balance in mouthfeel and, 83
  balancing neutral, coating, and
    contracting with, 66–67
  matching wines and food and,
    247–248
Dubourdieu, D., 278

E

Eggs
  cooking, 126–127
  at different temperatures, 100

Elegance, matching wines and food and, 264, 265
Emulsifiers, 96
Emulsion, 148–149
Energy, 127
En papillotte, 137–138
Enteric nervous system, 284
Enzymes, 141, 184, 186
    beer brewing and, 193, 194
    proteins and, 101
    in saliva, 24
Escoffier, August, 137
Escoffier, Auguste, 6, 138
Espresso, 67, 178, 191
Evolution, 111–112
Excitotoxins, 92
Expectancy, liking food and, 287
Expectation, 165
Extrinsic components of flavor, 278–280, 287–289
Extrinsic factors, 48–50

F

Fat replacements, 97
Fats. *See also* Lipids
    properties of, 95–98
    texture of, 38
Fatty acids, 95
Females, variation in flavor registration, 29
Fermentation. *See also* Vinification
    malolactic, 216–217
    secondary products of, 214
Fernel, Jean François, 14, 66, 76, 78
Fetus, 285
Filiform papillae, 25–26
Fish, 103
Flavor(s)
    defined, 3–4, 22
    enhancement or suppression of, 44–45
    expectancy and, 287
    fat as carrier of, 96

interrelation of tasting, perception, and, 10
    omission of, 160–161
    registration of, 13
    role of basic, 84–93
    as source of pleasure, 4–5
    synergy of, 161
    technical and emotional sides of, 10
    universal flavor factors, 14–16, 64–72
    words used to describe, 61–63
Flavor composition, 159–170
Flavor intensity. *See* Flavor richness
Flavor noise, 15, 71
Flavor profiles, 79–84, 159–160
    as dynamic, 82–84
    matching wines and food and, 234–236
Flavor registration, 10–12
Flavor richness (flavor intensity), 15–16, 68–70, 168
    of beer, 266
    matching wines and food and, 235, 236
    mouthfeel and, 74
    proteins and, 103
    successive contrast and, 46
    temperature and, 83
Flavor style 1, 76
Flavor style 2, 78
Flavor style 3, 78
Flavor style 4, 78–79
Flavor style 5, 76, 78
Flavor style 6, 78
Flavor style 7, 78
Flavor style 8, 78–79
Flavor styles, 16, 73–74
    basic characteristics of, 75
    description of, 76–79
    food appreciation and liking and, 291–296, 298
    matching wines and food and, 243–245

theorem of, 293–294
typical examples of, 77, 81
Flavor styles cube, 73–76, 79, 105
Flavor type (fresh or ripe flavor tones),
69–70
Flavour (magazine), 9
Fluency, 255
Foliate papillae, 25–28
Food appreciation and liking, 164,
271–302
behavioral model of food choice,
298–300
cross-modal interaction of the
senses, 50, 51, 278–279
development of preferences, 285–286
expectancy and, 287
external influences on liking,
287–289
flavor styles, 291–296
guestronomy, 300–302
need and, 280–282
neural pathways, 273–275
pleasure and, 282–283
stimulus-response relationship,
296–297
synthesizing neural information,
276–277
Food labels, 292–293
Food molecules, 93–104
carbohydrates, 98–100
lipids (fats and oils), 95–98
proteins, 100–104
water/moisture, 93–95
Foodpairing, 163. *See also* Matching
foods and beverages; Matching
wines and food
Food preparation. *See* Preparation
techniques
Food quality, 169
Freezing, 94
vegetables, 129
Fresh, 71
Fresh flavor style, 78
Fresh flavor tones, 15, 69–70

Fructose, 84, 85, 99
Fruits, 153–155
climacteric, 124
nonclimacteric, 124
seasonality of, 156–159
Frying
deep, 136
pressure, 144
shallow, 136
stir, 131–133
Frying/pan frying/stir frying (Bao
technique), 131–132
Full flavor style, 78
Functional magnetic resonance imaging
(fMRI), 14
Fungal infections, winemaking and, 206
Fungiform papillae, 25, 26, 28–30, 39,
40

G

Garum, 91
Gastrointestinal system (the gut),
284–285
Gastronomy
changing rules of, 6
defined, 1–2
evolutionary roots of, 111
as science, 10–12
Gelatin, 107
Gelling agents, 99, 142
Ghrelin, 284
Glocker, E. F., 61–62
Glossopharyngeal nerve, 27
Glucose, 84, 85
Glutamic acid, 31, 65, 79, 90, 91. *See
also* Monosodium glutamate
flavor richness and, 103
in wine, 215
Glycerol, 214
Goosefoot family, 151
G-protein-coupled receptors (GPCRs),
274–275
G-proteins, 274

Grapes and grape growing, 119, 202–210. *See also* Vinification
  acidity and sugar content, 207–208
  agricultural choices and practices, 204–206
  harvesting, 207–210
  machine or hand picking, 209–210
  phenolic ripeness of grapes, 207
  skins of, 210–211
  terroir, climate, and grape variety, 202–204
Green tea, 185
Grilling, 134
Guanylic acid (GMP), 91, 140, 185, 215
Guar gum, 107
Guestronomy, 300–302
Guide Culinaire (Escoffier), 6
Gustation, 22
Gustatory factors, 22
Gustatory system, 25–32
  interactions with olfactory system, 50–51
  interactions with trigeminal system, 51
The gut (gastrointestinal system), 284–285
Gymnemic acid, 45

H

Hams, cured or smoked, 139
Hänig, David, 14, 30
Harmony, 290
  matching wines and food and, 255–256
Hearing, 48, 50
Heat transfer, 127
Hedonic contrast, 46
Herbs and spices, 149–150
High fructose corn syrup (HFCS), 85
Hippocrates, 295
Holism, 294
Hops, 194, 197
Hospitality, 257

Hot compounds, 40
Hotel schools, 16–17
Hydrochloric acid (HCl), 31
Hydrocolloids, 148
Hydrophilic substances, 24, 47, 95, 99, 245, 246
Hydrophobic substances, 24, 47, 65, 95, 99
Hydroxypropyl methylcellulose, 106

I

Ice cream, 66
Ikeda, Kikunae, 14, 91
Infants, 285–286
Ingredients, basic qualities of, 115–125
  climate and location, 118–119
  production methods, 121–123
  varietals, cultivars, and breeds, 119–120
Inosinic acid (IMP), 91, 140, 185, 215
Integrity, 290
International Organization for Standardization (ISO), 22
Intrinsic components of flavor, 48–50
Ion-channel group (IC), 274
Ion channels, 40, 78, 97, 273
Irritants, 24, 40, 41, 44, 78

J

Jiro Dreams of Sushi (film), 161
Journal of Sensory Studies, 22

K

The Kitchen as Laboratory, 9
Korsmeyer, Carolyn, 9, 291

L

Labeling
  foods and wines, 292–293
  sugar consumption and, 86

Lactic acid, 87–88
Lager beers, 196
Lees, 214–216
Lemon juice, 141
Leptin, 85, 282, 284
Lettuce family, 153
Liking foods. *See* Food appreciation and
    liking
Lipids (fats and oils). *See also* Fats; Oils
    carry-over effects, 47
    properties of, 95–98
Liquamen, 91
Liquid nitrogen, 141, 142
Location, 118–119
Low-fat foods (low-fat diet), 97–98

M

McGee, Harold, 8, 114
Maillard reaction, 50, 103, 128,
        131–133, 144, 162, 171, 190,
        197, 266
Making Sense of Taste (Korsmeyer), 9,
        291
Malic acid, matching wines and food
        and, 240–241
Malolactic fermentation (MLF),
        216–217
Malt, 193
Marinating, 140
Marketing, of wines, 223–225
Mastication, 23
Matching beer and foods, 265–267
Matching foods and beverages, 231–267.
        See also Matching beer and
        foods; Matching wines and
        food
Matching wines and food, 79, 80,
        83–84
    acidity and, 88, 239–241
    cheese, 250–254
    cultural context and, 257–258
    dominant factor and, 235–236
    expensive wines, 260–261

flavor styles and descriptions,
        243–245
gastronomic reflections, 254–265
gustatory effects in, 237–245
harmony and contrast, 255–256
marriages made in heaven, 261–262
mouthfeel and, 245–254
new guidelines, 234–236
old guidelines, 233–234
reactions of basic flavors on wine
        types, 240
regional foods and wines, 262–263
rice test, 237, 242
robust and elegant dishes, 263–265
sugar and, 85
sweetness and, 242–243
Meat
    carving, 137
    influence of preparation, 172–173
    production methods and, 121–122
    ripeness and aging, 124–125
    searing, 137
Mechanoreceptors, 39, 78
Methylcellulose, 106
Microcrystalline cellulose, 106
Microvilli, 25
Microwave cooking, 144–146
Milk, 96–97
Mineral waters, 178–182
    interaction with wine and food,
        180–182
*Modernist Cuisine: The Art and Science of
        Cooking* (Myhrvold), 8
Moisture, 94. *See also* Water
    importance of, 137
Molecules, food, 93–104
    carbohydrates, 98–100
    lipids (fats and oils), 95–98
    proteins, 100–104
    water/moisture, 93–95
Moller, Per, 9
Monosaccharides, 84
Monosodium glutamate (MSG), 31, 92,
        167

Mouthfeel, 15, 22, 37–38, 64–67
  balancing neutral/dry, coating, and
    contracting, 66–67
  coating, 64–67
  combination of hard and soft
    textures, 167–168
  contracting, 64–67
  matching wines and food and, 234,
    235, 245–254
  from neutral to dry, 66
  proteins and, 102
  starch and, 99
  in theoretical model, 72–74
Mushrooms, 155
Music in restaurants, 279
Myhrvold, Nathan, 8

N

Name, 165
Nasal irritants, 24, 40, 41, 44, 78
Neural pathways, 273–275
Neurogastronomy (Shepherd), 22–23, 285
*Neurogastronomy: How the Brain Creates
    Flavor and Why It Matters*
    (Zuker), 9
Neutral flavor style, 76
Neutral substances, 66
  balancing dry, coating, and
    contracting with, 66–67
Nightshade family, 152
Noble rot (*Botrytis cinerea*), 65, 206, 210
Nociceptors, 39–41, 78
Nonclimacteric fruits, 124
Non-tasters, 29, 30, 40
Nutritionism, 295

O

Oak barrels, vinification and, 217–220
Odorants, 23, 24, 33–35
Oils
  for deep-frying, 136
  flavor intensity and, 47

properties of, 95–98
Older people, 27, 36–37, 166, 286
Olfactory cells, 33
Olfactory elements of flavor, 22
Olfactory system, 12, 33–37
  age and health and, 36–37
  odor composition and preferences,
    35–36
Omega-3 fatty acids, 98
Omega-6 fatty acids, 98
*On Food and Cooking: The Science and
    Lore of the Kitchen* (McGee),
    8, 114
Onion family, 151
Ono, Jiro, 161
Oolong tea, 186
Opioid peptides, 283
Oral irritants, 24, 40, 41, 44, 78
Oranges, 62–63
Orbitofrontal cortex, 255, 275, 280, 282,
    283
Organic agriculture, 122
Orthonasal smell, 33
Oxalic acid, 151
Oxidation, winemaking and, 222–223

P

Palatability, 163–165
  aesthetics of gastronomy and,
    289–291
  fats and, 97
  objectified, 169–170
  preparation techniques and, 113–114
  smell and, 165–166
Papillae, 25–31
  personal differences in, 27–28
Papillotte, 137–138
Pauillac wine, 262
Pavlov, Ivan, 296
Pectin, 100, 107
Peptides, 215, 282, 283
Perception of flavor, 4, 9, 10
Phenolic ripeness of grapes, 207

Phenylthiocarbamide (PTC) test, 29
pH of waters, 180
A Physical Essay of the Senses (Le Cat), 276
Physiologie du Goût (Brillat-Savarin), 5
Pickling, 140–141
Pineapple, influence of preparation,
     173–174
Pleasure, 282–283
    flavor as source of, 4–5
Poaching, 129–130
Polarity of water, 94
Pollan, Michael, 122
Polysaccharides, 41, 84, 99, 100
Port wines, 250
Potassium chloride (KCl), 87
Potatoes, influence of preparation, 173
Prenatal food experiences, 285
Preparation techniques (cooking), 17,
     128–136
    altitude and pressure, 143–144
    cold cooking, 138–140
    cooking and blanching, 128–129
    deep-frying and shallow frying, 136
    drying, pickling, marinating,
     140–141
    essence of, 125–128
    flavor tones and, 69
    flavor type and, 16
    frying/pan frying/stir frying (bao
     technique), 131–132
    grilling and barbecuing, 134
    influence of, 172–174
    mouthfeel and, 65
    poaching, 129–130
    purposes of, 113–115
    roasting/baking, 133
    searing, papillotte, clay pot, croûte,
     sous-vide, 137–138
    simmering, braising, or stewing, 135
    steaming, 130–131
    time and temperature variables,
     126–128
    very cold cooking, 141–143
Presentation, 162–163, 165, 287

Preservation, 113
Pressure, 143–144
Pressure cooker, 144
Pressure frying, 144
Price of wines, 224
Primary tastes, 13–14
    problems and misconceptions,
     30–32
    receptors for, 8–9
Production methods, flavor and,
     121–123
Professional tasters, amateur *vs.,* 51–52
PROP (6-n-propylthiouracil), 29, 30
Proteins
    coating, 65
    properties of, 100–104
Prutkin, J., 29, 39
Pu-erh tea, 186
Puff pastry, 67
Pungent flavor style, 78

Q

Quinine, 31, 32, 45, 89, 90

R

Radiance, 290
Radiation, 127
Raw foods, 113
Receptors (receptor neurons), 14, 273,
     274, 284
    for basic tastes, 8–9
Reductionism, 294, 295, 303
*Red Wine with Fish. The New Art of
     Matching Wine with Food*
     (Rosengarten and Wesson),
     233
Registration of flavor, 10–12
Residual sugar, 215
    matching wines and food and, 242
Retronasal smell, 33
Rice test, 237, 242
Ripe flavor tones, 15–16, 69–70

Ripeness, 71
   influence on flavor, 123–125
   phenolic, of grapes, 210–211
Roasting, 133
Robust flavor style, 76, 78
Römertopf, 138
Rooibos, 187
Round flavor style, 78

S

Saccharides. *See also* Carbohydrates
Saliva, 24
   matching wines and food and,
      247–248
   neutral substances and, 66
Salt, 45, 61–62
   brining or curing in, 139
   characteristics of, 86–87
   flavor intensity and, 68
   as hygroscopic, 162
   liking, 281
Saltiness
   matching wines and food and, 238
   of mineral waters, 182
   umami and, 167
Satiety, 282–284
Sauces, 146–149
   matching wines and food and, 246
Sautéing/stir frying (Chao technique),
      132–133
Scoring wine and food matching,
      79–82
Searing, 137–138
Sensory specific satiety, 283
Sensory thresholds, 42–43
Serotonin, 284
Shallow frying, 136
Shepherd, Gordon, 22–23, 285
Simmering, 135
"Six ages of flavor," 285–286
Size, 143, 161–162
Skinner, B. F., 296
Slow eating, 283

Smell. *See also* Odorants; Olfactory
   system
   liking and disliking food, 281
   other senses "needed" by, 51
   palatability and, 165–166
   taste and, 12–14
Smoking, 139, 140
Sodium chloride (NaCl), 31
Soil, 119
Solvent, water as, 95
Somatosensory system. *See* Trigeminal
   system
Sour, 87–88
Sous-vide, 137, 138
Sparkling wines, 216
   matching food and, 246–247, 252
Spherification, 142
Spiciness (spicy foods), 44
   beer and, 267
   matching wines and food and, 239,
      249–250
Spinal nerves, 276
Spring water, 178–180
Starch, 107
   in sauces, 148
Steaming, 130–131
Stenden University, 9
Stewing, 135
Stinging compounds, 40
Stir frying, 131–133
Stock, 129, 135, 147
   sauces and, 147
Successive contrast, 46
Succinic acid, 214
Sucrose, 45, 84
Sugar(s), 84–85
   suppression of flavor and, 45
Super-tasters, 29, 30, 40
Suppression, of flavor, 44–45
Sushi, 161
Sweeteners, artificial, 85–86
Sweetness, 31, 32
   of beers, 198, 267
   characteristics of, 84–86

of fruits, 154
matching wines and food and, 238,
    242–243
suppression of flavor and, 45
temperature and, 82
Sympathetic nervous system (SNS), 282
Synergy of flavors, 161
Synesthesia, 50

T

Tannins
    matching wines and food and, 248
    vinification and, 210, 211, 216
Tastants, 23–24
Taste. *See also* Gustation
    defined, 2–3
    etymology of, 38
    smell and, 12–14
Taste buds, 25–29
    turnover rate of, 27
Taste pores, 25
Taste receptor cells (TRC), 25
Tasters, professional *vs.* amateur, 51–52
Tastes, basic (primary), 13–14
    problems and misconceptions, 30–32
    receptors for, 8–9
Tasting
    the brain and, 272–277
    flavor profile and, 79
    interrelation of flavor, perception,
        and, 10, 11
    as intricate process, 22–23
    what happens in the mouth, 23–25
Tea
    aromatized, 187
    black, 186
    chemicals in, 183–184
    flavor styles of, 184–187
    green, 185
    oolong, 186
    Pu-erh, 186
    uses of, 187
    water for making, 183–187

Temperature(s), 126–128
    for beer brewing, 196
    of eggs, 100
    flavor perception and, 51
    flavor profiles and, 82–83
    mouthfeel and, 64
    poaching, 129
    for serving wines, 225–226
    of waters, 182
        for making tea, 185
    winemaking and, 213–214
Terroir of grapes, 202–203
Textures, 22, 23
    combination of hard and soft,
        167–168
    defined, 38, 39
    mouthfeel and, 38
Theanine, 183
Theoretical model, 72–76
Thermoreceptors, 39
Thickening and gelling agents, in
        processed food, 106–107
This, Hervé, 9
Thomas Aquinas, St., 290
Time, cooking, 126–128
    influence of altitude and pressure,
        143–144
Time intensity, sensory thresholds and,
        42–43
Tongue, 276, 277
    zones of, 14
Tongue map, 30
Total dissolved solids (TDS), 178
Transduction, 23
Trans fats, 97
Trigeminal elements of flavor, 22
    olfactory system and, 34
Trigeminal system, 23, 24, 27, 37–42
    astringency, 41–42
    interactions between gustatory
        system and, 51
    nociceptors and, 40–41
    receptors of, 38–40
Triglycerides, 95

U

Umami, 25, 31, 45
  characteristics of, 90–92
  flavor intensity and, 68
  flavor styles 4 and 8 and, 78–79
  palatability and, 166–167
  wines and, 215–216
Universal flavor factors, 14–16, 64–72
  mouthfeel, 64–67

V

Varietals (cultivars), 119–121
Vegetables, 150–153
  bitterness in, 90
  seasonality of, 156–159
  size of, 162
Ve-Tsin, 92
Vichy Catalan, 180
Vinaigrette, 67
Vinegar, 141
Vineyards, 119, 262
Vinification (winemaking), 210–220
  after, 220–226
  blending, 220–221
  further aging, 222–223
  lees and, 214–216
  malolactic fermentation, 216–217
  phenolic ripeness and maturity of
    skins, 210–211
  temperature and, 213–214
  wooden barrels and, 217–220
  yeasts and, 211–214
Viscosity, 78
Vision, liking and disliking food, 281
Von Liebig, Justus, 137

W

Wageningen University, 9
Wardle, J., 292
Water
  for beer, 194

grape growing and, 205–206
  mineral or spring, 178–182
  properties of, 93–95
  soft, 182
  soil and, 119
  for tea, 183
Whitefish, influence of preparation, 172
Wine and food matching. *See* Matching
    wines and food
Wine labels, 292–293
Winemaking. *See* Vinification
Wines, 198–227. *See also* Grapes and
    grape growing; Vinification
  aging of, 204, 216, 218, 221–223
  appellation of, 200
  astringency of, 41
  balancing different elements in, 67
  blending, 220–221
  Brettanomyces and, 43
  coating, 65
  decanting, 226
  differences between beer and,
    226–227
  environment influences on
    perception of, 279
  flavor factors, 201–202
  flavor intensity, 68, 69
  flavor profiles, 82
  marketing of, 223–225
  matching foods and. See Matching
    wines and food
  matching with flavor tones, 70
  mineral waters and, 180–12
  mouthfeel of, 201, 207, 210–212,
    214–216, 218
  price of, 224
  sauces and, 149
  serving, 225–226
  serving temperature of, 82–83
  sparkling, 216
  words used to describe, 63–64
Wooden barrels, vinification and,
    217–220
Words, to describe flavor, 61–63

X

Xanthan gum, 107, 148

Y

Yeast extract, 92
Yeasts, 140
    beer brewing and, 195
    winemaking and, 211–214

Yogurt, perception of the sweetness of,
    278–279

Z

Zinc deficiency, 27
Zuker, Charles, 8–9

Printed in the United States
by Baker & Taylor Publisher Services